GW00372544

Aircraft Alive

Aircraft Alive

Aviation and Air Traffic for Enthusiasts

Chris McAllister

B T Batsford Ltd, London

To my pilots, Brian and Tim,
who showed me how it all worked

Line illustrations by Anne Fox and Val Caldwell

First published 1980

ISBN 0 7134 1914 8
Filmset in Monophoto Apollo by
Servis Filmsetting Ltd, Manchester
Printed in Great Britain by
Butler & Tanner Ltd
Frome Somerset
for the publishers, B T Batsford Ltd,
4 Fitzhardinge Street, London W1H 0AH

Contents

Figure 1 *Reproduced by courtesy of British Air*

1-Aviation Enthusiasts

We British are an air-minded nation. Many of us are fascinated by flying and by aeroplanes and will turn our heads at the sound of a plane approaching. We cannot help looking to see where the plane is, what type it is and what it might be doing. Inside many people there is an aviation enthusiast trying to get out and it is for these that this book has been written.

This is a book for people who like watching aircraft, especially aircraft in flight. It is about where to find aircraft, where they go when they take off and how they get to where they are going. It is about the bits and pieces which help them on their way; charts, flightpaths, radio, airspace and navaids. *Aircraft Alive* is about planes which have their wheels in the air.

I prefer the term 'aviation enthusiast' or 'air watcher' to 'plane spotter' with its slightly derisory, schoolboyish connotations. An intelligent interest in aviation is not confined to any one sex, age group or social bracket and this you will see for yourself on the spectator terraces of any airport, or if you visit an aviation museum or flying display.

The great thing about being an aviation enthusiast is that you can pursue your interest at your pace and in your own way. There is no special organisation you must join, equipment you should purchase, venue you have to attend or rules you must follow. The only requirement is the obvious one that you should be keen on aircraft. Some people can watch all the aircraft they need from their lounge window, if they happen to live near a busy flightpath. You could fit in a bit of plane spotting when driving across country – for example on holiday – if you detour or stop off to visit an airfield. There are museums, displays, air fairs and RAF Open Days up and down the country which you could attend. Perhaps you like reading about flying or watching aviation programmes on television. If you want to talk about flying, you need the company of some like-minded friends, so you might join a club or society of aviation enthusiasts. From plane spotting you can branch out in several directions, such as listening-in although this is illegal at the moment on air-band radio. You can specialise, if you wish, in military flying as opposed to civil, or vice versa. You can try aircraft photography, a rewarding hobby which does not require any very special equipment. You can join an aircraft preservation society and help to build and restore historic aircraft, or you can roam the country looking for traces of vanished wartime airfields or crashed bombers. Ultimately, if you can afford it, you can book a set of flying lessons and obtain your Private Pilot's Licence (PPL). Quite literally, the sky is the limit!

This book won't take you quite that far, but it should be an invaluable handbook on some of the more modest rungs of the aviation enthusiast's ladder. There is a need for books which explain the technicalities of flying to the non-specialist enthusiast layman. Many aircraft books in the shops are good on aircraft and engines, but give very little information on navigation, navaids, radio and avionics, and the arrangements governing airspace and Air Traffic Control, particularly as they apply to the UK. It is these gaps especially which this little book hopes to fill.

Finding Aircraft

Aircraft can be seen on the ground at any airport. You can visit military airfields too, but only on special occasions such as Open Days. Aviation museums are also worth a visit. The exhibits may be largely static, but many of the types on display are modern and you can walk right up to them and inspect or photograph them as closely as you wish. However, if you want to watch aircraft in flight, you need to know about flightpaths.

It is perfectly possible to live close to a major airport and not hear or see many aircraft. The reason is that aircraft cannot approach to land at or climb away from an airfield except along routes which line up with the runways. The approach and departure flightpaths are extensions of the runway centreline. If you have a large-scale map of the airfield and its surroundings, such as the Ordnance Survey One Inch, or alternatively the 1:50,000 series, you can plot these flightpath lines for yourself. Draw a *thin* pencil line down the middle of the runway and extend it out into the countryside for several miles in both directions. Then, if you go and park yourself along this line, you should be able to observe planes flying more or less directly overhead. Successive aircraft follow in the wake of each other with a precision which may make you wonder if there were not rails in the sky. In the case of approach flightpaths to large civil airports the rails are the radio beams of the Instrument Landing System (ILS) which are projected out along the extended runway centreline, and slope up from the ground at an angle of 3°. Receivers aboard the aircraft can detect and lock on to these beams from as far away as twenty miles out, riding them down out of the sky onto the runway, losing height at the rate of 300 feet per mile. This method of getting aircraft onto a runway is capable of great precision. Watch aircraft approaching a busy airport – in the case of Heathrow at peak periods they come in at a rate of one every 2 minutes –

14
14
16
16

Alitalia
JAT
YUGOSLAV AIRLINES
YU-AKB

Figure 2 Aircraft alive. Bulldog elementary trainers from RAF Leeming over the White Horse of Kilburn, near Sutton Bank in North Yorkshire. *MOD*
Figure 3 Aviation enthusiasts at Heathrow

and each aircraft seems to slice through exactly the same piece of sky at exactly the same height as its predecessor.

The closer you get to the airfield boundary the lower the planes pass above your head. Here you can take exciting photographs without the need for expensive equipment such as telephoto lenses, but you should try to use a fast shutter speed.

Meanwhile, at the other end of the runway, planes are taking off. If you choose to watch here you will find it very noisy. You may also find it disappointing, as modern aircraft are soon up and away, climbing steeply to disappear into the distance or into cloud before you can get a proper look at them. Departing aircraft follow closely-regulated flightpaths, and although these are not usually extensions of the runway, you may find them worth studying.

Civil aircraft travel from A to B along specially designated corridors called airways, which are marked out by radio beacons. These are the main roads of the sky, and on a clear day they can show up as parallel rows of vapour trails crossing overhead at anything up to 40,000 feet as the world's airliners go about their business in the skies over Britain. With the help of air-band radios some aviation enthusiasts specialise in identifying these overflights, even though the planes are almost unrecognisable pale wraiths high in the blue overhead. You may find a use for a telescope and tripod here.

Military aircraft in flight are less regimented in their movements than their civil counterparts. Training sorties at both high and low levels are flown in UK airspace. The Navy, RAF and USAF lay on a generous programme of flying displays throughout the country each summer, and there are places where the enthusiast can get to watch some aspects of military flying training, such as bombing practice.

Then there are the hundreds of small fixed-wing aircraft and helicopters scattered throughout the country engaged in what is called General Aviation business; air taxi and charter operations, private and business flying, tuition and training, crop spraying, etc. When convenient to do so, these use the airways, but often they fly quite low in order to keep clear of their bigger brothers manoeuvring under air traffic control. These Aztecs, Apaches and Cherokees get to be quite numerous around 1000 feet, a slice of the sky which has come to be dubbed 'Indian Territory' by the pilots of larger aircraft.

First, therefore, find your flightpaths. In some ways aircraft watching is a bit like, say, bird watching. Bird watchers begin by studying the haunts and habitats of the species they want to find. For example, you will not find golden eagles nesting on Hampstead Heath, and for that matter you will not find Jaguars from Lossiemouth using it for low-level training runs! It might however be a convenient place from which to watch civil traffic in and out of Heathrow and Northolt. Very roughly, it could be said that different parts of Britain are used for different kinds of flying.

The sketch-maps in this book give a very simplified picture of the aeronautical geography of the UK, suggesting as they do, rigid subdivisions of airspace between the civil and military interests. In practice any such arrangements are by no means rigid, and the picture is in any case a three-dimensional one, difficult to depict in any detail on a small sketch-map. We advise the enthusiast to consult larger scale aeronautical charts for his information rather than rely on the small sketch-maps in this book. These cannot show much detail and cannot reproduce the full colour and tone ranges of the original charts. They are intended merely as illustrations to the text.

You can buy aeronautical charts brand-new from a specialist supplier. However, if you belong to an aviation society, flying club or school, ATC unit etc., or if you know any pilots you should be able to obtain discarded charts. Pilots are required to use up-to-date editions of topographical and aeronautical radio charts, so when a new edition is published, the old ones are discarded. You may be able to arrange for one or two charts to come your way.

Many flightpaths are defined by radio beacons and other navigation facilities (navaids). Aviation enthusiasts who are into the geography of flightpaths may be encouraged to look for these navaids and to try to understand how they work. Most navaids are simple radio transmitters, but some are weird and spectacular installations such as Doppler VORs. For technical reasons they are mostly situated in open country, but some are able to work satisfactorily in built-up areas. Beacons and other navaids are dealt with fully later in the book.

Aircraft Recognition

The ability to recognise aircraft types easily and quickly was of vital importance during the Battle of Britain and indeed throughout two World Wars. The Royal Observer Corps, who did the official plane spotting in those days, trained using three-view silhouettes of allied and enemy aircraft. Thus the three-view silhouette has come down to us as the basic tool for discerning between aircraft types, and is used in almost all present-day recognition manuals, such as

The Observer's Book of Aircraft, which is about the best known. However, things have changed since World War Two. In particular it is becoming very difficult to tell many aircraft types apart; they all look so very alike these days, e.g. T-tail airliners, and three-view silhouette drawings don't always bring out the differences very well.

You should aim to be able to recognise aircraft types when seen a long distance away, as misty silhouettes or as momentary glimpses. You need good eyesight (or a pair of field glasses) and the kind of brain that registers shapes accurately. Above all, you need practice. Some types are distinctive, such as a Harrier, Concorde or a Boeing 747, but can you tell a Boeing 707 and a DC-8 apart? The latter has a droopy nose and a different slant to the fin. On approach to land, the L-1101 Tristar, the DC-10 and the A-300 Airbus look very similar, but with practice you can begin to spot the differences. Other aircraft groups which are difficult to distinguish are light aircraft, especially twins, and business jets.

Rather than relying on three-view silhouettes, you may find it better to visit an airfield and study the aircraft types in the flesh. In this way, your brain will begin to register distinctive features which consciously you haven't even noticed! Photographs are a great help, and it is a good idea to compile a scrapbook of pictures from magazines, showing different aircraft types from a variety of different angles.

Yet another idea worth trying is to build models of types which look alike, such as the Trident and the Boeing 727, both of which are available in plastic kit form. Study the completed models and compare them when seen from various angles and the differences will begin to stand out.

A useful training suggestion I once heard was for enthusiasts to get together and try to identify aircraft types from photographs presented by means of a tachistoscope. This is a kind of viewer device used by experimental psychologists to present pictures and images, displayed on postcards, for split-second duration and/or under low illumination. You may be able to borrow a tachistoscope from a local school or college; the dim, fleeting images it can present are very typical of the way many aircraft are seen in the field.

With the help of some practice, then, you should begin to find aircraft recognition a fascinating challenge. You may even lose your fear of T-tails, and soon be able confidently to distinguish between BAC 1-11s, Tridents, DC-9s, Boeing 727s, Fokker F-28s, VC-10s, IL-62s, HS-125s, Falcons of various sizes, Gulfstreams, YAK 42s, Tu 134s, and 154s!

Plane Spotting

The narrower meaning of this term is the activity of collecting aircraft registration markings (reggies). With many aviation enthusiasts this is their major pastime. Special books are published for plane-

Figure 4 General aviation: the beautiful Piper Navajo Chieftain is much in demand for air taxi work and is popular with pilots and passengers alike. *Norman Edwards Associates*

spotters which give markings in alphabetical or numerical order, with details of type, operator, etc. An excellent example is *Civil Aircraft Markings* published by Ian Allan Ltd and updated each year. It is available from most aviation bookstalls. This lists civil aircraft currently in service in the UK and Ireland, and those in airline service in other countries, i.e. 98% of the civil aircraft you are likely to encounter in the UK. For rarer birds, military serials, etc., you may need more specialist publications, or help from members of an aviation society.

These spotters' books are useful when learning to identify aircraft types or type variants. Thus if you see G-AXCP and G-AVMU parked next to each other, the book will tell you that although both are BAC One-Elevens, one is a 401 and the other is a 510 version. Try comparing the number of passenger windows forward of the wing.

Air Radio Watching

Small portable radios which can be tuned to the VHF aircraft frequencies between 108 and 136 MHz are now readily available and inexpensive. Many aviation enthusiasts, and in particular plane spotters, enjoy listening-in on these radios to the exchanges between pilots and controllers. They can also be used as an aid to plane spotting itself. Tuned to the airport's approach frequency, they can be used to listen out for anything interesting or unusual on its way in, and spotters who live near airports or ATC transmitters can use these radios from their own homes. Reception conditions vary from place to place. The aircraft can be heard easily enough, but in order to hear the controllers, it helps to be within line-of-sight of a transmitter, i.e. in open country, on a hilltop or a tall building. These radios are probably of most interest to civil aircraft enthusiasts, because the air-band frequencies (108–136 MHz) to which they can be tuned are those used by civil aircraft. Military aircraft use a range of frequencies in or near the UHF band (e.g. 300 MHz) and as yet nobody has mass-produced a receiver which can be tuned to military aircraft frequencies.

The big snag is that the use of these radios is illegal, though this is not often realised. (See Chapter Four.) Aviation enthusiasts would be well-advised to observe the law and not to use these radios, or if they must listen, to do so discreetly. Too often one sees plane spotters in airport lounges with their radios parked beside them. Then the voice of the pilot or the controller will come very loudly from several sets simultaneously. This is asking for trouble. Airport authorities do not like to persecute aircraft enthusiasts, but they must often wonder whether they ought to enforce the law, and ask the airport police to move these spotters elsewhere. It would be better if the spotters were to listen silently through earphones, at least when indoors on airport property. If you can't be good, be careful.

Join a Club

Any activity is more enjoyable if it is pursued in the company of friends with the same interests. You will enjoy aircraft more and make many new friends if you join one of the many aviation societies which are based here and there throughout the country. They organise meetings, film shows, etc., and run coach trips to displays and other events in distant places. Some have their own clubrooms, and in one case at least these are on airport premises. Members can meet to exchange ideas, information, photographs, etc. Some societies are active in preserving and restoring aircraft. Of these the Duxford Aviation Society is perhaps the best known. Others such as the Merseyside Aviation Society specialise in publishing, and the booklets, maps and leaflets they produce are sought after by aviation enthusiasts all over the UK.

Although there are many clubs and societies, there is as yet no national organisation to which they are all affiliated. Such a body would not only represent the interests of aviation enthusiasts to national and government concerns such as the CAA and the British Airports Authority, but it could furnish the addresses of the secretaries of local aviation societies to anybody who enquired. As things are, about the best way to find out if there is an aviation society or spotters' club in your locality is to make enquiries at the local airport. If the information desk cannot help you, then you are reduced to asking any plane spotters you can see around.

Aircraft Museums

The big advantage of aviation museums is that you can walk around quite close to the aircraft on display. Not all the aircraft housed in museums are antiques either. There are Concorde prototypes or pre-production models at Duxford and Yeovilton, and scattered around other museums you will find Comets and Viscounts, Hunters and Buccaneers. Many museum aircraft are still active and are flown on special occasions. Some of the more important collections are as follows:

Figure 5 Aviation museums: young and old together on a wet day at Duxford

Figure 6 Displays and Open Days are good for aircraft close-ups. *Keith Price*

Figure 7 'Finals'. Position yourself near the runway threshold for good studies of aircraft approaching to land. *Keith Price*

Imperial War Museum, Lambeth Road, London SE1. Other London collections are at the Science Museum, Kensington and the RAF Museum, Hendon. All are worth a visit.

The Imperial War Museum's Reserve Collection is preserved at Duxford, near Cambridge, with the able assistance of the Duxford Aviation Society which does much of the restoration work on a voluntary basis.

Based at Old Warden, near Biggleswade, is the Shuttleworth Collection, containing a large number of preserved aircraft of World War One vintage. Many of these are capable of being flown, and flying displays are held on the last Sunday of each month, March to October.

The Fleet Air Arm Museum is based at RNAS Yeovilton and contains examples of almost every type of British naval aircraft. The RAF Museum is based at Hendon, but there is an RAF Aerospace Museum at Cosford, near Wolverhampton and a collection of 'historic' aircraft at RAF St. Athan, although these can only be seen by prior arrangement. In Scotland there is the Museum of Flight at East Fortune, near North Berwick, and the Strathallan Collection at Auchterarder in Perthshire. Both of these are well-stocked.

Smaller museums include the British Rotorcraft Museum at Weston-super-Mare, the Cornwall Aeronautical Park at Helston, the Historic Aircraft Museum at Southend, the Midland Air Museum at Coventry Airport, the Mosquito Aircraft Museum at London Colney near St. Alban's, the Newark Air Museum, Torbay Aircraft Museum, Thorpe Water Park, near Egham in Surrey, and the Wales Aircraft Museum at Rhoose Airport.

Aircraft Photography

If you have a camera you will probably want to use it to take pictures of aircraft. The results are equally rewarding in colour or in black and white. However, before you rush out to spend hundreds of pounds on specialist equipment such as telephoto lenses, here are

AIR
A
B

Figure 8 Aircraft photography: try shooting into the light for softer, more pleasing results.
Figure 9 Aircraft photography: use of an orange or red filter will show up the clouds

Figure 10 Aircraft photography: camouflage patterns (and the tones of RAF roundels) separate out when you use filters. *MOD*
Figure 11 Aircraft photography: the wing and tail of the nearby aircraft frame those in the background. *Manchester International Airport Authority*

498

Figure 12 Aircraft photography: night scenes like this are difficult but challenging. *Paul Francis Photography*
Figure 13 Aircraft photography: go for the detail! *MOD*

some simple tips which will work even if you don't have a very elaborate camera.

Parked aircraft ought to be easy enough to photograph. There may be a problem getting close enough to them, or getting yourself into the right position, or getting rid of lighting standards, jetties, flights of boarding steps and other airport paraphernalia. The problem is particularly bad at Heathrow, but less serious at some of the smaller airports. At an aviation museum, there are none of these problems – you can usually walk as close to the aircraft as you wish and photograph points of detail.

Shots of aircraft taxiing, lining up, taking off and landing can sometimes be taken through the airport boundary fence, if this comes near the runway. You can poke your camera lens through the wire mesh, but I know of one enthusiast who carries around in his car a small set of lightweight kitchen steps. Standing on these he is tall enough to get his camera over the wire!

If you do decide to obtain a telephoto, then a 200 mm. lens (fixed or zoom) on a 35 mm. camera is a good arrangement. You could fit a much longer lens, such as 400 mm., but this monster will definitely need a tripod, and on long shots in summer you could have problems with haze. Tripods are useful, but there isn't always room to use one, especially at air displays. Use a reasonably fast shutter speed, even for static shots, e.g. 1/125th sec., otherwise you may find some of your more rare and difficult poses marred by camera-shake.

If you use black-and-white film, try experimenting with lighting and filters to get more subtle effects. Orange and yellow filters will emphasise clouds in a blue sky. They will also separate the tones in some military aircraft camouflage schemes, and even make the RAF roundels visible! Do not always shoot with the sun directly behind you, but use it to model the detailed parts of the aircraft. With confidence you will learn to shoot *into* the light, increasing the exposure by one or two stops to bring out the detail in the shadows.

CAUTION
CAUTION
478

Figure 14 Aircraft photography: try for the more interesting angles. *MOD*

Aeronautical Geography

The study of aeronautical charts and flight guides is full of interest for aviation enthusiasts. These documents show in detail the location and layout of flightpaths, airways, airspace arrangements, and the beacons and other navigation aids which mark out on the ground the geography of the air.

You can easily collect aeronautical charts. The various types of chart and the sources from which they may be purchased are explained in detail in Chapter Three. However, as we have mentioned earlier, professional aviators discard their old editions of charts whenever new ones are published, and you may find that you can easily waylay these secondhand charts, if you know anyone who works at an airport. To an enthusiast, aeronautical charts are attractive possessions, full of absorbing interest.

Navaids and Avionics

Shown on these charts are the navigation aids which mark the airways and flightpaths. Some are simple radio transmitters – non-directional beacons they are called – little brick huts with tall aerial poles standing alongside. Others are the much more sophisticated navigational beacons known as VORs (short for VHF Omni Range). In some places a newer and even more sophisticated navaid known as a Doppler VOR is being installed. The military equivalent of a VOR is TACAN, and then there is DME, Decca, and the group of installations which make up ILS – the Instrument Landing System. An intelligent enthusiast will want to know something about these – how they work, what they look like, and above all where they are. A full treatment of electronics, radio, etc. is beyond the scope of this book, but we hope the explanations given in Chapters Four and Five will tempt readers to delve a little deeper. There are more than enough aviation books on the market which deal with airframes, engines and performance figures, for example, but precious few which make any attempt at all to explain avionics. There is a definite need for a book which explains at a not-too-technical level what these little brick huts are all about.

2-Airfields in Action

Scattered throughout Britain there are over 1000 places which have served as airfields at one time or another. One legacy of the last war is hundreds of military airfields which are now silent and disused. The Spitfires, Fortresses and Lancasters have now gone and the farmers have returned, although the concrete runways may remain, having resisted efforts to break them up. Of the airfields in use today, some are civil, some military, some large, some small.

Civil Airfields

These vary in size and in the nature of the facilities they provide, from Heathrow at one end of the scale, down to the smallest weekend airfield, perhaps without radio and with grass strips instead of concrete runways. The more important airfields can provide customs facilities, if not round-the-clock, then at least during the working day. Any aircraft departing for abroad or arriving from abroad must be cleared at one of these.

Customs Aerodromes

As soon as an arriving aircraft stops it is boarded by a Customs Officer, and until clearance is complete no one may disembark, if he does he must not leave the Customs area. There are roughly 50 designated customs aerodromes dotted throughout the UK. Most of these are equipped with some kind of bad-weather landing aid, such as radar or the ILS (Instrument Landing System). If the main runway is long enough and of a suitable specification it will be available to large jet aircraft. The boom in the package-holiday (inclusive tour) trade in recent years now means that large jets operate from airports all over the country. Not only is it more convenient for passengers to use the regional airports, but the landing fees and apron and hangar charges are usually more reasonable, and these savings can be passed on to the tour passengers. Some British airports are run by the British Airports Authority, but a number are operated by local authorities, and the remainder by aerospace companies, private enterprise, etc.

Access and Viewing

Airports exist first and foremost to serve the travelling public. The relatives and friends of these, who come to meet and greet them, are second in an airport's order of priorities. Thirdly there are those who visit airports to look at the aircraft. Not all of these are fully-avowed aviation enthusiasts or plane spotters, but members of the public who perhaps still cannot convince themselves that anything weighing 350 tons can heave itself into the air, then fly to the other side of the globe without stopping, at a speed close to 600 m.p.h. A modern airliner is more than just a magic carpet; it represents an almost incredible technical achievement in performance, comfort and safety right on the very edge of the possible.

Most airports make provision for plane watchers to look at aircraft, and this provision varies. Depending on the airport you may find balconies, piers, car-parks, and even warm lounges with snack-bar facilities. There may or may not be an admission charge. Some have booklets on sale giving the arrival and departure schedules; this saves much leafing through airline timetables.

For example, take London's Heathrow Airport; Daddy of them all with more planes coming and going, and a greater variety of them than anywhere else in the country. It ought to be, and is, a Mecca for all aviation enthusiasts. Some have been known to stay for days, and even go without meals, in order not to miss a single aircraft movement. In the central area, reached via the tunnel, among the main passenger terminals you will find the main spectator facilities on the roof of the Queen's Building, which is connected by a walkway to the roof of Terminal 2 also. These spectator terraces are open to the public from 10.30 a.m. each morning. Depending on the security situation, you may be able to station yourself on the roof of one of the multi-storey car-parks, which are good vantage points. Back again at ground level you could explore the perimeter road. If you get on at the northern interchange and travel west, you will find that there is a spectator car-park.

Continuing round the perimeter road, you will find that you have to keep driving. Double yellow lines forbid you to stop. All the same, you still get close-up glimpses of aircraft landing and taking off and of such installations as the ILS localiser transmitters. You will pass the cargo and general aviation handling areas, where you may glimpse the occasional biz-jet rarity. Finally you come to Hounslow, where British Airways have their Heathrow headquarters and maintenance facilities. Beware of large aircraft crossing the road! Perhaps not on the perimeter itself, but all round Heathrow there are vantage points which have been sought out by plane spotters over the years. The best advice I can give you is that you go and seek out these nooks and crannies for yourself. Don't forget that there is a sensitive security situation involving British airports at most times. Should you be challenged remember that discretion is the better part of valour.

The above pen-picture of Heathrow can be scaled down and made to fit most other airports. Besides the official spectator facilities, there will be other vantage points if you explore around the boundary or perimeter. Some of these are access to crash-gates which you must not block with your car. An Ordnance Survey One Inch or 1:50,000 map is invaluable in getting to know the geography of your local airport and its surroundings, including where the flightpaths are situated, but even a road map will be helpful in finding vantage points.

Airports and Enthusiasts

For a variety of reasons airport staff are friendly and helpful towards aviation enthusiasts. They are useful allies in an age when the public has started to become more hostile towards airports and aircraft in some cases. The plight of noise-sensitive people who live near large airports is very real, and it is not just aircraft noise, but the increased traffic along local roads that is seen by many people to have become a hazard and a nuisance. Airports are in need of friends these days and aviation enthusiasts of all ages count as useful friends and allies. However, this friendship and trust must be safeguarded. Airport staff have a job to do and their main responsibility is for the comfort and safety of passengers. Anyone who travels by air knows that it sometimes involves long and infuriating waits at airports, and the peace of mind of somebody who can't wait to get away on his Spanish holiday is not likely to be helped by the sight and sound of hordes of plane spotters with radios, rampaging through the passenger lounges. Take care not to become a nuisance to passengers or to pester airport officials, and remember, the use of air-band radios is illegal. If you wish to listen to yours on airport premises it is best to use an earphone.

In recent years, however, we have seen chaos and crisis descend on many British airports, particularly in the summer months. Airport and airline staff have group-by-group begun industrial action, causing delays and hold-ups. Air traffic control staff working-to-rule, whether in Britain or in Europe, causes an artificial rationing of airspace, and some flights have been delayed for days. In such conditions the airport terminals are filled to overflowing, and are more than overstretched trying to cater for their stranded clientele. If you find that as a mere airwatcher you are unwelcome inside the airport, please be patient and do come back again when the crisis has passed.

Furthermore, there is the question of security. In recent years airports have become targets for terrorist activities, hijackings, shoot-outs etc. International terrorism knows no frontiers, and our own airports, normally open to all and sundry, can easily become involved. If an incident should take place, or if one is threatened, airports may be closed to the non-travelling public for some weeks. Plane spotters are turned away and multi-storey car parks are cordoned-off. Unless you are booked on a flight you will not be allowed inside the terminal. These restrictions are irksome while they are in force, but they are usually lifted after a few weeks.

Figure 15 A French-registered Beech Super King Air arrives at a British airport where it is met by a marshaller and a customs officer

Figure 16 Inclusive Tour aircraft operating out of Manchester. An Airtug is all ready to push back the One-Eleven in the foreground, and in the background a Boeing 737 is already taxiing

HEATHROW AIRPORT: Facts and Figures

Runways 10R/28L 12,000 feet long, 300 feet wide

10L/28R 12,802 feet long, 300 feet wide.

05/23 7733 feet long, 300 feet wide.

Helicopter Pads (a) On Northern perimeter just to the west of tunnel entrance.

(b) On General Aviation apron

Car Parks (a) Longterm parking adjacent to Northern Perimeter Road.

(b) Multi-storey short term car parks in central terminal area.

Spectator Viewing Areas (a) In Central area, on roof of Queen's Building, from 10.30 a.m. until dusk.

(b) Around Perimeter Road – varies.

Radio Frequencies Tower 118.7, 121.0 MHz

Ground Movements 121.9, 118.5 MHz

Approach Control 119.2, 119.5, 120.4, 127.55 MHz

Control Zone Transit 119.9 MHz

ATIS Automatic Terminal Information Service (Departure) 121.85 MHz

ATIS Arrival Information from Bovingdon VOR 112.3 and Biggin 115.1 MHz

Engine Start Clearance 121.7 MHz

Pushback/Taxi Clearance 121.9 MHz

Approach and Landing Aids VOR/DME London (LON) situated half a mile north of the airfield perimeter at OS Grid Ref 066776 Frequency 113.6 MHz

Instrument Landing System (ILS)

Runway	*Frequency MHz*	*Callsign*	*DME Channel*	*Outer Marker* (Distance from runway threshold—nautical miles)	*Middle Marker*	*Approach Beacon* (NDB) Frequency kHz	Callsign
28R	110.3	I-RR	40	3.7	0.9	—	—
28L	109.5	I-LL	32	3.8	0.8	334	OE
10R	109.5	I-BB	32	3.8	0.7	—	—
10L	110.3	I-AA	40	3.4	0.8	389.5	OW
23	110.7	I-CC	—	4.1	0.7	334	NE

Figure 17 The layout of a runway as seen by an approaching pilot

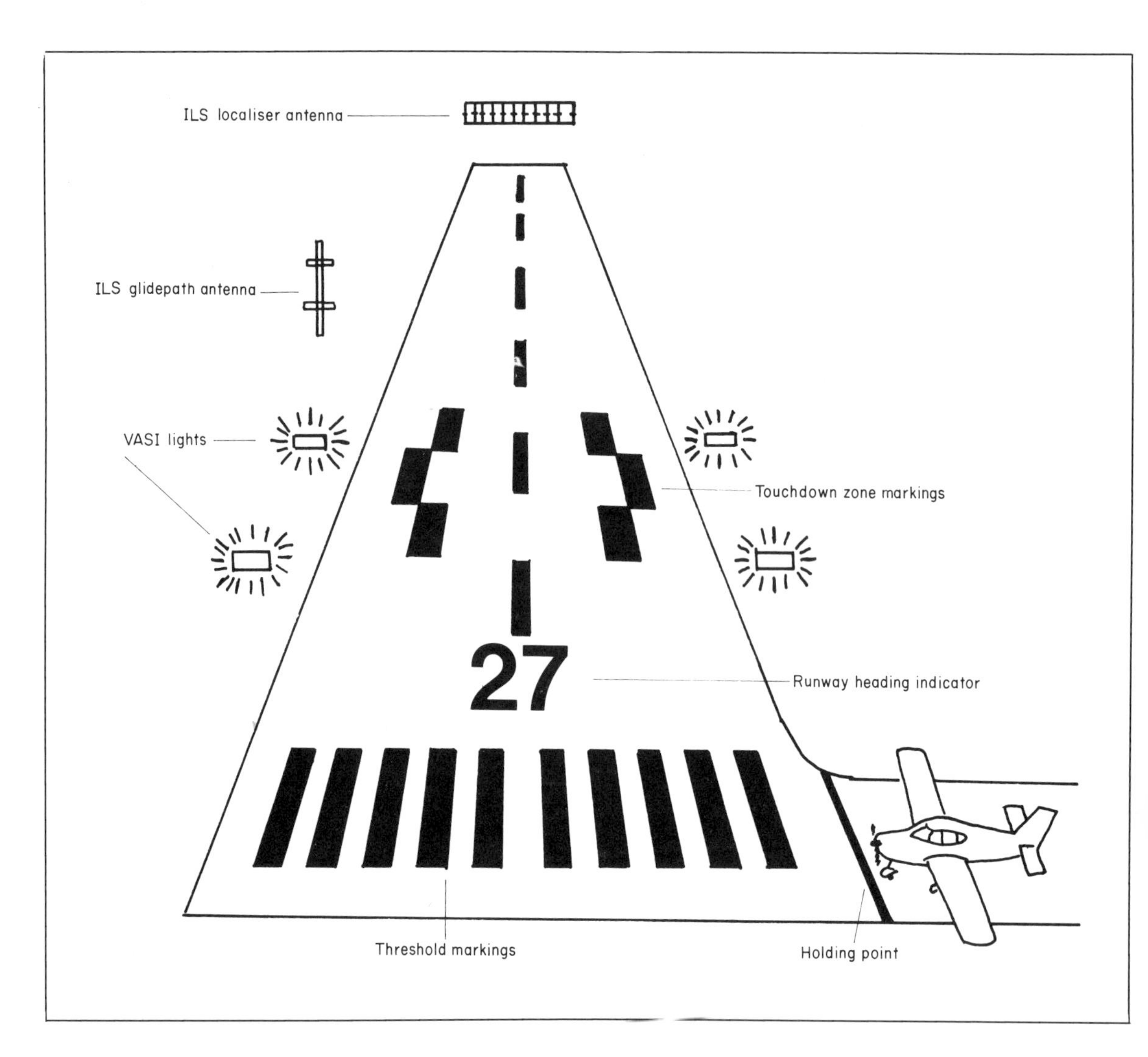

Runways

Only light aircraft can get away with operating from grass. Anything fast or heavy ploughs up the surface, especially in wet weather, and hard, bumpy ridges are left when the field dries out. A system of permanent runways in concrete or asphalt, each about a mile in length and laid usually to form a triangle took shape at most British civil or military airfields in the 1930s and 1940s. This triangle still underlies the runway layout at many airfields today. The original three runways can still be found but at least one will have been lengthened to cope with today's aircraft types. This will now be the main runway, fitted with sophisticated lighting and landing aids, and will handle almost all the airfield's traffic. The two remaining pre-war runways are probably little more than taxiways today, but the one upgraded runway and its extended centreline will be the focus of all the action.

It is worth noting the system used to refer to runways. It is based on the direction in which they run – their heading in relation to magnetic north. This is rounded to the nearest 10 degrees and the final zero omitted. Thus for a runway which is aligned 239° magnetic, we round to 240, omit the zero and we get Runway 24. This will be painted in large white letters at the approach end of the runway. As planes can also approach from the other end, this will have the magnetic heading painted there too, in this case the reciprocal of 24 – which is 06 (add or subtract 18). When you refer to these runways you say 'Runway Two Four' and 'Runway Zero Six'.

Heathrow has three main runways, two of them in

Figure 18 Enthusiast's map of London's Heathrow Airport

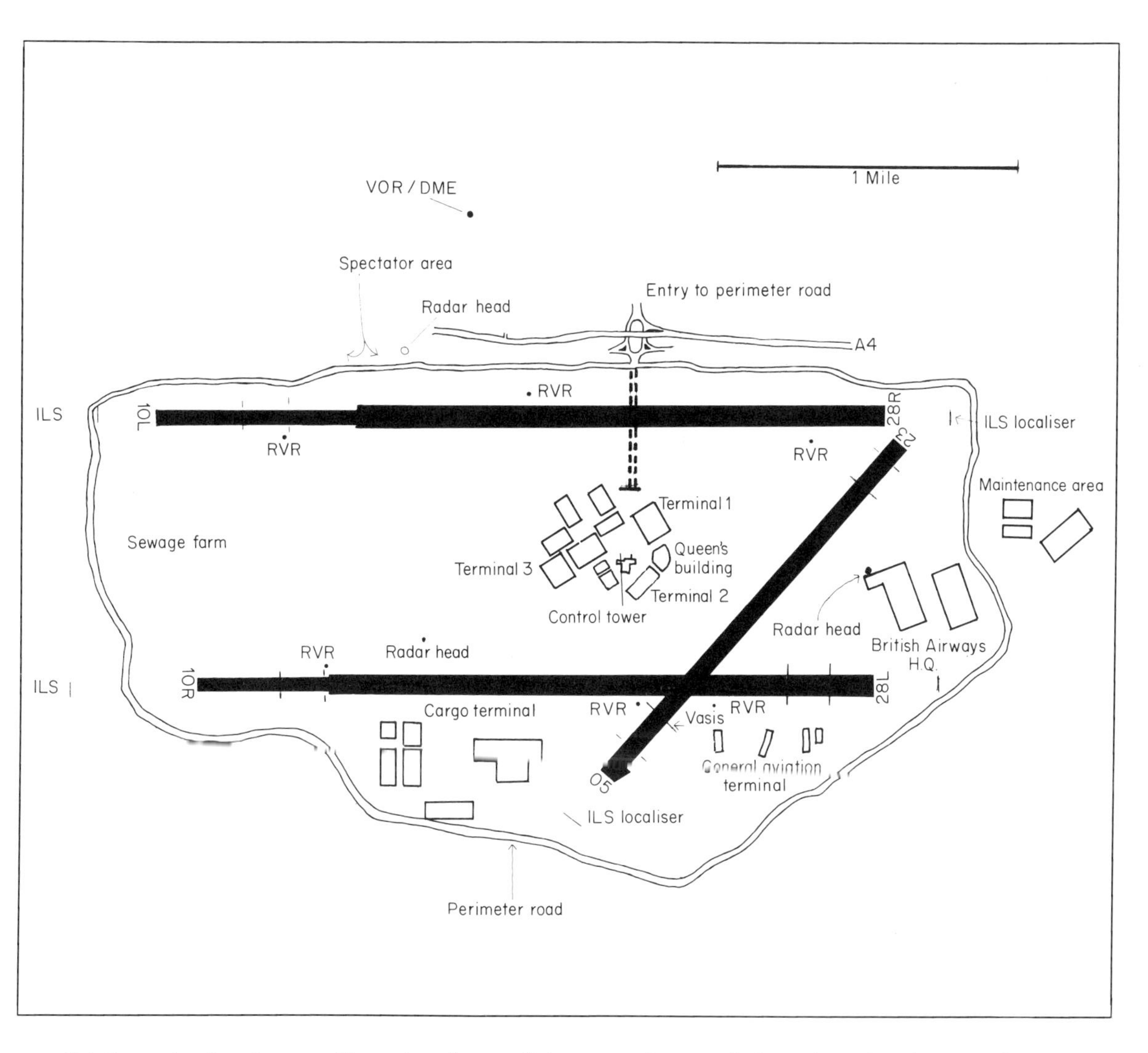

parallel. From the sketch you will see that the parallel runways are 28R/10L and 28L/10R. A third runway is aligned 23 and 05. Normally two runways are in use, one for landing, the other for take-off. For example in a strong south-westerly wind, Runway 23 would be best for landing, and take-offs might use Two Eight Right. When there is little or no wind, westerly operations are preferred for noise control reasons. Planes do not begin to take off in an easterly direction until a headwind of at least 3–4 knots has established itself. Similar preferential runway systems operate at other airfields.

Runway Markings

As a pilot approaches the runway to land, he will see the *threshold* marked with a set of broad white stripes. It is safe to touch down here, but the normal practice is to aim for the specially-marked *touchdown zone* 1000 feet or so farther along the runway. Other markings show the runway centreline, the runway heading, holding-points, taxiways, distance-to-run, etc.

The Control Tower

The Tower is the glasshouse, usually with sloping double-glazed windows, placed so as to have a commanding view of the airfield as a whole. The controllers in the Tower are responsible for ensuring that the runway is used by one aircraft at a time (or one formation at a time), giving each aircraft clearance to take off and land and controlling the movements of aircraft and vehicles on the ground. This is possible with the help of radio, and each control tower has its

Figure 19 Belfast by night. Lighting and VASIs show clearly on the approach to Aldergrove's Runway 26. *Brian Wrathall*

Figure 20 Concorde touches down on Heathrow's Runway 28L. In the background is the control tower dominated by the housing containing the airfield radar scanner. *British Airways*

Figure 21 London Gatwick. An ideally-served airport with immediate access to rail and motorway. The threshold and touchdown zone of Runway 26 can be seen clearly. *BAA Gatwick*

own frequency or channel, normally VHF or UHF. Small airfields without radio have to rely on signal boards, Verey pistols and other outmoded systems.

After radio, the air traffic controller's biggest ally is radar. This can be used to 'see' aircraft which are still too far away to be visible to the eye, and it is convenient for the radar to be operated from a semi-darkened room in or near the tower. The radar controllers have their own separate radio channels, which they use to communicate with the pilots of the aircraft which appear as blips of light on their screens.

Air traffic control officers are highly trained and experienced. One feature of their training is that they should learn to fly and qualify at least to Private Pilot standard. The RAF trains its controllers at Shawbury, and the Civil Aviation Authority has its training unit at Hurn, near Bournemouth. Some controllers are trained by private undertakings, such as International Aeradio, based at Southall. If you are thinking of a career in Air Traffic Control and would like to know about entry requirements and selection procedures, you should contact these organisations. (Civil Aviation Authority, *CAA House, 45–59 Kingsway, London WC2B 6TE;* RAF Inspectorate of Recruiting, *Ministry of Defence, Government Buildings, Stanmore, Middlesex, HA7 4PZ*; International Aeradio Limited, *Hayes Road, Southall, Middlesex.*)

Airfield Installations

Around the airfield you will notice a number of fixtures and fittings and will want to know what they are.

First the *lighting*. An airfield is a spectacular place at night, but the lights are often left switched on during the day. They are an added reassurance to pilots if there is any cloud or haze about. *The approach lights* in particular can be designed as foglamps, capable of piercing through thin cloud and murk. They enable a pilot approaching on instruments to line up visually with the runway seconds earlier than he otherwise would.

Two banks of red lights close to the touchdown zone show as white lights if you fly over them. They are designed so that a pilot approaching at the correct angle (3°) will see the far set as red, the near set as white. These are the *VASIs – Visual Approach Slope Indicators*. One white, one red, you're on the slope, both red, you're too low; both white, you're too high.

Transmissometers are little yellow-painted robots shaped like toadstools which measure the visibility. The RVR (Runway Visual Range), as it is called, must be monitored closely; if it is less than 600 metres then the airfield may have to be closed. The old method of measuring visibility was to station a man in a draughty RVR tower counting the runway lights, and the RAF still does this. However at most civil airports the automatic photoelectric transmissometers have

British airways

Figure 22 Instrument Landing System localiser at Castle Donington
Figure 23 A Plessey AR 15/2 radar head at RAF Valley. *Plessey Radar Limited*

taken over, signalling the RVR continually to the tower. Other weather instruments such as thermometers, wind vanes and anemometers are connected to the control tower in the same way.

The main radio transmitting and receiving aerials are located not in the control tower, but on masts and towers some distance away. To cut down interference, receiving and transmitting aerials are in different places.

A radar scanner (radar head) is easily recognised. Different airfield radars provide different information, and some work better in rain than others. At Heathrow there is a special short-wavelength radar for mapping aircraft and vehicles on the ground, and this is housed in the orange 'dustbin' feature on top of the control tower.

Navigation and Landing Aids

In or above cloud aircraft must rely on radio aids to navigate, to find their destination and to make a safe approach until they are below the cloud and in sight of the runway. If you are interested in this side of aviation you will want to know what these ground-based aids look like, how they operate, and where they are located. Later chapters in this book will cover much of this detail, but while you are in or around the airfield you might encounter one or more of the following:

The aerial of a *radio beacon* is either a tall pole or a wire stretched between two poles. Aircraft can home in on the beacon and thus find the airfield, but modern practice is to position the radio beacons along the approach flightpath, which is where the pilot really wants to be. This is where the beacons are at Heathrow and Gatwick, for example. There is also at Heathrow a special kind of beacon called a *VOR/DME* (see Chapter Five). This looks like a large dustbin sitting on the roof of a low hut and can be seen to the north of the A4 just across some fields. This beacon is mainly used for aligning departure routes after take-off. It transmits on a frequency of 113.6 MHz, and on an air-band radio the Morse callsign LON (·-·· --- -·) can be heard.

There may also be some kind of *Direction Finder* installed on the airfield. One type consists of a hut with a circular flat roof.

Most of the transmitters connected with the *ILS (Instrument Landing System)* are positioned on or near the runway itself. The *Localiser* is the odd-looking fence-like array of aerials at the far end of the runway. This emits a pattern of radio signals which guide the pilot along the extended runway centreline from as far out as 25 miles or more, and he is guided down the 3° glideslope from another array of aerials positioned alongside the runway near the touchdown zone. This is where you can see the *Glidepath* transmitter as it is called. More bits and pieces complete the ILS installation. There are *marker beacons* out along the approach flightpath, and *monitor receivers* dotted around the airfield. These look like VHF television aerials and they check the stability of the signal pattern being transmitted by the ILS.

Take-off and Departure

Let us imagine we are watching a jet airliner preparing to depart. It is due to leave in about half an hour or so. The flight-crew are already aboard going over the numerous routine checks designed to ensure that every piece of equipment is functioning correctly. A flight-plan will have been filed with Air Traffic Control and the captain will have checked the latest weather bulletins and briefed his crew on any out-of-the-ordinary aspects of the flight. He has worked out how much fuel he needs to take on board; enough for the flight, plus 45 minutes' reserves, plus enough for a diversion to an alternate destination should that be necessary. Fuel is important in any pilot's calculations; in a long-range jet it is crucial. More than half the take-off weight of a Boeing 747 could be in liquid form. It is a factor in most of the calculations prior to departure, such as how much runway will be needed for take-off and how soon cruising altitude can be reached. The huge fuel load taken aboard before a long-distance flight may not leave the aircraft with much of a performance margin.

The engines have not yet been started. The noise you hear comes from the *Auxiliary Power Unit (APU)* mounted in the tail. This supplies power for the aircraft's systems while the main engines with their associated pumps, generators, etc., are still shut down. When the aircraft is ready the passengers are marshalled aboard, and as soon as the runway is likely to become free Tower gives the captain clearance to start the engines. When the engines are running there are more routine checks. An overalled engineer near the nosewheel is still in telephone contact with the pilot via a plug-in headset. A special tractor, called an airtug, is ready to push the aircraft back until it is clear to move ahead under its own power.

After pushback the engine note gets louder and the plane starts to taxi forwards, steering with its nosewheel. It follows a system of taxiways until it reaches the entrance to the main runway, where it may have to stop and hold for a few minutes. When cleared, it turns onto the runway, straightens up and usually wastes no

British airways
G-BDDA

AVIOGENEX

Figure 24 Dyce Airport, Aberdeen, in winter. *A Shell Photograph*

Figure 25 Soviet-built aircraft are regular visitors to British airports. A Tupolev TU-134 collects tour passengers bound for Yugoslavia

Figure 26 The flames of the reheat show clearly as with an ear-splitting roar this Phantom from RAF Germany reaches for the sky at Wildenrath. *MOD*

time at all before opening up and starting the take-off roll. The engines now have the job of accelerating this large heavy machine to take-off speed, about 150 m.p.h., and there is only the length of the runway in which to do it. Similarly, if after this point the pilot decides to abandon the take-off for any reason, there will not be much runway left in which to stop.

The captain will have worked out in advance the maximum speed at which he can abandon the take-off should he need to, e.g. in the event of engine failure, and still bring the aircraft to a safe halt before the end of the runway. This speed, called V_1, depends on a number of factors, i.e. take-off weight, condition of the runway, etc. and can be worked out from tables published in the Flight Manual for the aircraft type. Up to and including V_1 the nose is held down. The wheels must still be allowed to take the full weight of the aircraft in case the brakes have to be used suddenly. Beyond V_1 it will not be possible to stop safely and the plane is committed to take off. It continues to accelerate until take-off speed is reached. This is V_R – the airspeed at which the pilot pulls back on the control column to '*rotate*' the aircraft. The nose lifts into the take-off attitude and a second later the plane is airborne, most probably climbing away from the runway at a steep angle. The brakes are used to stop the wheels spinning, the landing gear is retracted and the wheel-well doors close with the loud clang which always seems to startle first-time airline passengers. The flight is under way.

An old-type piston-engined aircraft would have gained height slowly; you may have seen films of wartime bombers struggling to get airborne with their massive loads of bombs, but a modern jet or turboprop transport has power to spare and can leap skyward at a steep angle, especially if it is only on a short flight and has not uplifted a full load of fuel. If you are watching from the outbound end of the runway the aircraft may be several hundred feet above your head by the time it reaches you. As a passenger too, you will have noticed how rapidly the scenery shrinks to doll's house size as you climb away from the airfield. At about 500 feet a

Figure 27 An A 300 B2 Airbus 'rotates' its nose upward into the take-off attitude. *British Aerospace*
Figure 28 Airborne: all 350 tonnes of 747. *British Airways*

turn may be commenced. Then at about 1000 feet a surprising thing happens; the engines are throttled back.

Yes. The roar of powerful jet engines which first pushed you back into your seat while still on the runway now dwindles to a placid murmur. The pilot has reduced power out of regard for the Noise Abatement regulations and the people who have to live and work underneath the flightpath. There is however no need to worry. There is still enough power on to keep the aircraft climbing steadily with the help of flaps and other high-lift devices. After a few minutes of climbing like this the noise-sensitive areas are left well below and behind and full power can be restored. The flaps can be wound in as the airspeed increases and the aircraft, now in 'clean' configuration, continues in an en-route climb to its cruising altitude, which in the case of subsonic jet aircraft is anything from 20,000 to 40,000 feet.

Aircraft Noise

Modern jet aircraft are extremely noisy, particularly during the take-off and climb, and it is the people who live and work under the flightpaths who suffer the most from the noise nuisance. People are variously affected. Many can accept any reasonable level of aircraft noise, but others on the contrary are very noise-sensitive. They can be greatly distressed by aircraft noise, even at low intensities. Soundproofing can help, but not if you must leave the windows open for ventilation. Schools, hospitals, elderly persons and mothers with young children are among the groups most likely to be affected by having to live under or close to airport flightpaths. Not everybody who lives near Heathrow is an aviation enthusiast. The more articulate members of the community can ring up and complain or form a pressure group, to try to get something done about it. But even if the level of noise from aircraft was substantially reduced, it is thought there would still be complaints.

The problem can be tackled at source. The Boeing

707 was always considered a very noisy machine, even to the point where it came to be regarded as the standard against which other planes were judged. Most present-generation short-haul jets are slightly less noisy, and the big wide-bodied jets such as the 747, L-1011, DC-10 and A 300 are very noticeably quieter than the Boeing 707. They even make a different sound. This is because they are powered by the large advanced-technology fanjets which were designed in the early Sixties, partly with a view to overcoming the noise problem. One way they do this is by directing a sheath of air backwards so that it envelops and quietens the hot exhaust gases from the jet efflux. Compare different types taking off and you can judge for yourself.

Noise can be measured using a sound-level meter. It is registered in Perceived Noise Decibels (PNdB). This is a logarithmic scale, and some examples of readings are:

Quiet country lane	20 PNdB
Whisper at 5 feet	34 PNdB
Conversation at room level	60 PNdB
Vacuum cleaner at 10 feet	70 PNdB
Heavy town traffic at kerbside	75 PNdB
Heavy lorry at 25 feet	90 PNdB
Boeing 707 taking off at 500 feet	110 PNdB

120 PNdB is the pain threshold for most people, however having to endure continuous noise of the order of 90 PNdB can cause loss of hearing over a period of time. This is the level of noise produced by many discotheques and rock groups. Borrow a noise-level meter and check for yourself.

Distance is the most important method of lessening aircraft noise. The law of inverse squares operates, and the energy of the noise at 1000 feet should be only a quarter of that at 500 feet. The standard *Noise-Abatement Take-off Procedures* recognise this principle. Use full power in order to make the take-off run as short as possible, then climb steeply to 1000 feet. Under favourable conditions a modern jet can be at 1000 feet before it reaches the airport boundary. This is Part One of the noise abatement procedures. Part Two now follows: at about 1000 feet the engines are throttled back to reduce noise (and power) and the plane then drifts relatively quietly over the noise-sensitive built-up areas below. There is still enough power to produce a steady and gradual climb, especially with a few degrees of flap out. A few miles farther on Part Two comes to an end; a few thousand feet more of altitude has been added, and fewer decibels are being registered down below. Full climbing power is now restored and the flaps can be wound in.

Not all pilots seem equally capable of following these restrictions. *Noise Monitoring Points*, set up on the ground below, usually about 3 miles out from the airfield, can help to catch out the 'cowboys'.

Figure 29 Minimum noise routeings, London/Heathrow. *Reproduced by kind permission of the Civil Aviation Authority*

Sanctions can be applied. These often take the form of a reduction in landing fees for aircraft which keep their take-off decibels below a certain figure.

For Air Traffic Control reasons, aircraft leaving most airfields are routed along special flightpaths. These *Standard Instrument Departure* tracks are defined by means of bearings taken from radio beacons, VORs, etc. Nowadays these SID tracks have been arranged so that near to the airfield they cross areas which are less built-up or less sensitive to noise. Thus we have *Minimum Noise Routeings,* which in the case of Heathrow are shown on the chart. Sometimes Air Traffic Control is able to spread the noise around a bit by using *Radar Departures.* Aircraft manoeuvre under control of ground radar, overflying other places, thus giving the Minimum Noise Routeings a rest for a while.

Approach and Landing

Now that we have said goodbye to our departing aircraft it is time to look towards the approach end of the runway to see if we can see another plane coming in to land. Most aircraft begin their final approach a long way out from the airfield – five miles or more – where they turn onto the localiser beam from the Instrument Landing System. They follow this towards the runway until they intercept the glidepath beam coming up from the runway at 3°, then they follow this electronic slope until the pilot can see the runway. The ILS is intended as a bad-weather landing aid, and can in fact be used when cloud is down to about 200 feet. It cannot be used to complete a landing as it is not reliable below this altitude, so in conditions of poor runway visibility, fog etc., the aircraft will have to divert unless it is fitted with some kind of automatic fully blind-landing system. However the ILS is commonly used in fine weather also. Pilots find an ILS approach easier than a fully visual landing, and controllers find that it is convenient to feed incoming aircraft one-by-one onto an ILS localiser beam knowing that they will then decelerate to a typical approach speed and land one

Figure 30 The Rolls Royce RB 211 high by-pass ratio turbofan engines fitted to this Tristar are much quieter than earlier jet engines. *British Airways*

after the other reasonably separated in time.

If the approaching aircraft has its landing lights on you will see it easily a long way out. Pilots use their landing lights in broad daylight so that they can be seen, either by the controllers in the Tower, or by pilots of other aircraft who may be manoeuvring to avoid them. If you see a short downward-curving plume of exhaust smoke this will be caused by a quick burst of throttle which the pilot used as he found himself momentarily below the correct ILS glidepath. In a jet however a quick burst needs to last quite a few seconds, as the engines do not accelerate quickly, neither does the aircraft flatten its approach too readily. The landing gear will be fully down and locked and the flaps will be out almost to full. A jet transport will be approaching at about 160 m.p.h., aiming to touch down at about 130 m.p.h. In a strong wind, particularly a strong crosswind, the final stages of the approach will require a good deal of piloting skill. Gusts give the aircraft a lot of extra lift one moment then take it away a moment later. Throttles need to be open in windy conditions just that bit extra in order to give more speed to manoeuvre. If there is also a strong crosswind component the aircraft will be approaching as though it is flying sideways! If you have never seen a large aircraft approaching in a crosswind, flying sideways and rolling from side-to-side, then you have a surprising and rather unnerving spectacle to look forward to next time there is bad weather at your local airfield!

As the aircraft crosses the threshold, the throttles can be closed and the pilot can straighten the plane out to line up with the runway. He has been losing 10 feet per second all the way down the 3° approach slope,

Figure 31 Checking noise levels. *Norman Edwards Associates*

and he must now flatten out – flare, as it is called – in order to touch down with hardly a bump at all. The flare will take him even farther along the runway before the wheels make contact, and it is necessary to get all three sets of wheels down firmly, taking all the aircraft's weight before the brakes can be used.

Watch the airbrakes. At touchdown they spring upwards from the top surface of the wing. Their function is to 'spoil' the airflow over the upper camber, thus preventing the wings from lifting. The alternative name for these powerful airbrakes is 'spoilers' or 'lift dumpers'. Full reverse thrust is now used, and finally the wheelbrakes, slowing the plane down to walking pace long before it reaches the end of the runway.

Designing hydraulic wheelbrakes for large heavy aircraft is a challenging problem. Brakes can easily overheat and start a fire if they are not used with caution. Cooler brakes would mean large heavy wheel assemblies, altogether too cumbersome, therefore anything which makes the job of the brakes easier is bound to prove immensely worthwhile, whether it be thrust-reversers, spoilers or parachutes.

Then there is the other friction surface, the runway itself. Many runways are alright in dry weather, but very slippery in the wet. The poor braking action is caused by aquaplaning on the layer of water which covers the runway surface. Specially-designed runways not only provide a slope and channels for the water to run away, but the surface is deliberately roughened, enabling the tyres to make contact even during a tropical rainstorm.

So, along you go to your local airfield and take it all in!

Figure 33 Panavia Tornado: how flaps and slats change the camber of the wing is shown clearly in this photograph. *British Aerospace*

Things to Do

1 Visit your local airfield and try to discover the best places from which to observe and photograph aircraft.

2 Obtain a large-scale Ordnance Map (One-inch or 1:50,000 will be ideal) of your local airfield and the surrounding area. Use thin pencil lines to continue the line of the main runways out across country. In this way you may be able to find other places at or near the airfield perimeter which give good views of aircraft approaching or taking off.

3 Try to find as many as possible of these on or near your local airfield;
(a) Radar heads (or 'scanners')
(b) Radio transmitting and receiving aerials
(c) VHF radio direction-finder (VDF) or any other direction finder
(d) Radio beacons or a VOR
(e) Anemometer (for measuring windspeed)
(f) Runway Visual Range (RVR) Transmissometers
(g) Visual Approach Slope Indicator (VASI) Lights
(h) ILS Localiser Transmitter
(i) ILS Glidepath Transmitter
(j) ILS Outer and Middle Markers
(k) Noise monitoring microphones (these are always well hidden).

4 If yours is a civil airfield, see if you can obtain a published schedule or timetable giving details of all regular take-offs and departures.

5 Watch for any strange or unusual aircraft visiting your local airfield. If you are a member of an aviation club or society, remember to tell your friends and colleagues about them too.

6 Use a stopwatch (or a watch with a sweep second hand) to time how long it takes different aircraft to accelerate to take-off speed.

7 If you can obtain the use of a noise-meter, set it up at the take-off end of the runway to compare the noise made by different types of jet aircraft, and under what conditions noise is the loudest. Try drawing noise contours for your airfield. Measure noise at the approach end of the runway also.

Figure 34 The 747: flaps and slats in front and triple-slotted flaps behind. *British Airways*

3-Aircraft in Flight

Getting to the Sharp End

As an aviation enthusiast you will itch to get airborne from time to time, but flying as a passenger in a large jet is bound to prove unsatisfactory. What you need to do instead is to get yourself up to the flight deck by flying in a small aircraft.

You can of course charter an aircraft if you are wealthy, or book yourself a set of flying lessons, but there may be no need to go to such extremes. Most small aircraft nowadays have seats for four or more, and many pilots have no objection to taking along the occasional supernumerary passenger provided he pays his share of the costs. It is in any case worth befriending private pilots and flying club members, because the least you can hope for is a chance to climb aboard a light aircraft to examine the layout of instruments and controls.

On the other hand, if you do decide to pay for a set of flying lessons you will need about 40 hours at about £20 per hour in order to qualify for your Private Pilot's Licence. This entitles you to fly your family and friends around but not, as they say, for hire or reward. You will need other qualifications too, such as the blind-flying (IMC) rating, night rating, and type ratings for other types of aircraft. You still pay £20 an hour to hire your aircraft, so many private pilots form a syndicate, buy an aircraft and share the costs. You can go on to qualify under your own steam for a multi-engine rating, and the coveted Instrument Rating, which allows private pilots to use controlled airspace, and with enough flying hours under your belt you could qualify as a Commercial Pilot. What has previously been your hobby, could now be your job.

For those who can pay their own way, the approved flying schools at Oxford, Carlisle and Perth can take a suitable trainee all the way to his Commercial Pilot's Licence in about 18 months. The course costs £20,000 so it is no surprise that most of the students at these schools are from abroad, sponsored by their national governments and airlines. British Airways trains pilots entirely at its own expense, and of course so does the RAF.

Figure 35 Flight deck of the Scottish Aviation Bulldog, used by the RAF as an elementary trainer. Note the Basic T formed by the main flight instruments. *MOD*
Figure 36 Flightdeck of Boeing 747. Slightly more complicated, but the Basic T is still there! *British Airways*

Controls and Instruments

Armed with a Private Pilot's Licence gained in a Piper Cherokee, nobody supposes that he would be able to step immediately into the Captain's seat of a Boeing 747, with its bristling flight deck full of levers and instruments, switches and warning lights in great profusion everywhere. However, both the Cherokee and the 747 are aeroplanes, and are flown in much the same way; pull the control column to bring the nose up, push it to force it down. Turn the wheel and the ailerons operate to control banking. Use the rudder pedals to help out in the turns. There are vast handling differences; the 747 is harder to fly than the Cherokee partly because it will not respond as quickly to the controls. There are of course so many extra systems which have to be mastered that it takes long training to cope with them all.

However, the light-plane pilot would feel at home sitting on the flight-deck of a Boeing 747 and staring at the instruments straight in front of him. These are the main flight instruments common to nearly all aircraft types and grouped (usually) in a fairly standardised way: i.e. in the form of a letter T. The vertical part of the T comprises the Artificial Horizon or its equivalent, with some form of compass below it. The left arm of the T is the speed indicator, and the right arm is the altimeter. These are the most important instruments, and there is room around the T to tuck in a few more. In order to prevent the pilot having to search across the panel for other instruments and possibly reading the wrong ones, planes such as the Boeing feed inputs from other sensors into the Basic T. Thus the so-called Compass is properly termed a *Horizontal Situation Indicator*, comprising inputs from magnetic and gyro compasses, radio compasses, VOR/ILS meters, etc. all in one instrument and distinguished by different pointers. The pilot has in front of him all he needs to steer a course or to set the autopilot to do so. The autopilot systems fitted to these large aircraft are sophisticated enough to take care of 90% of the flying, including standard manoeuvres.

Figure 37 The Basic T – the same arrangement as before – this time in an F-16 fighter. *Marconi Avionics Limited*

Figure 38 Flight-deck of HS 125 business jet. (See key.) *British Aerospace*

Figure 39 HS 125 Instrument Panel and Controls. (Key to illustration below.)

1 Combined Speed Indicator (ASI/Mach)
2 Attitude Director Indicator
3 Standby attitude indicator
4 Servo Altimeter
5 Horizontal Situation Indicator
6 Vertical Speed Indicator
7 Radio Altimeter
8 Angle of Attack Indicator
9 Elapsed Time Clock
10 Fuel contents
11 Fuel flow
12 Fuel consumed
13 L.P. spool RPM
14 Turbine temperature
15 H.P. spool RPM
16 Oil pressure and temperature
17 Brake pressure indicator
18 Cabin altitude control panel
19 Flap position indicator
20 Autopilot control panel
21 Flight guidance panel
22 Flight annunciators
23 Master Warning System Indicator
24 Master Warning Panel
25 Altitude alerting
26 ASR Weather Radar
27 VHF COMM 1
28 VHF COMM 2 (double tuner)
29 VHF NAV 1
30 VHF NAV 2
31 ADF control box
32 Radio Magnetic Indicator VOR/ADF
33 DME control box
34 DME display
35 Marker indicators
36 ATC transponder and marker receiver
37 VLF/Omega long-range navigation system
38 HF radio control panel
39 Microphone/intercom pushbutton
40 Attitude trim controls
41 Rudder trim
42 Power levers
43 Air brakes
44 Control column and wheel
45 Wheel brake lever
46 Landing gear indicator
47 Landing gear selector
48 Audio station control box
49 Emergency Oxygen
50 Cabin Voice Recorder microphone

RUDDER TRIM
AILERON TRIM

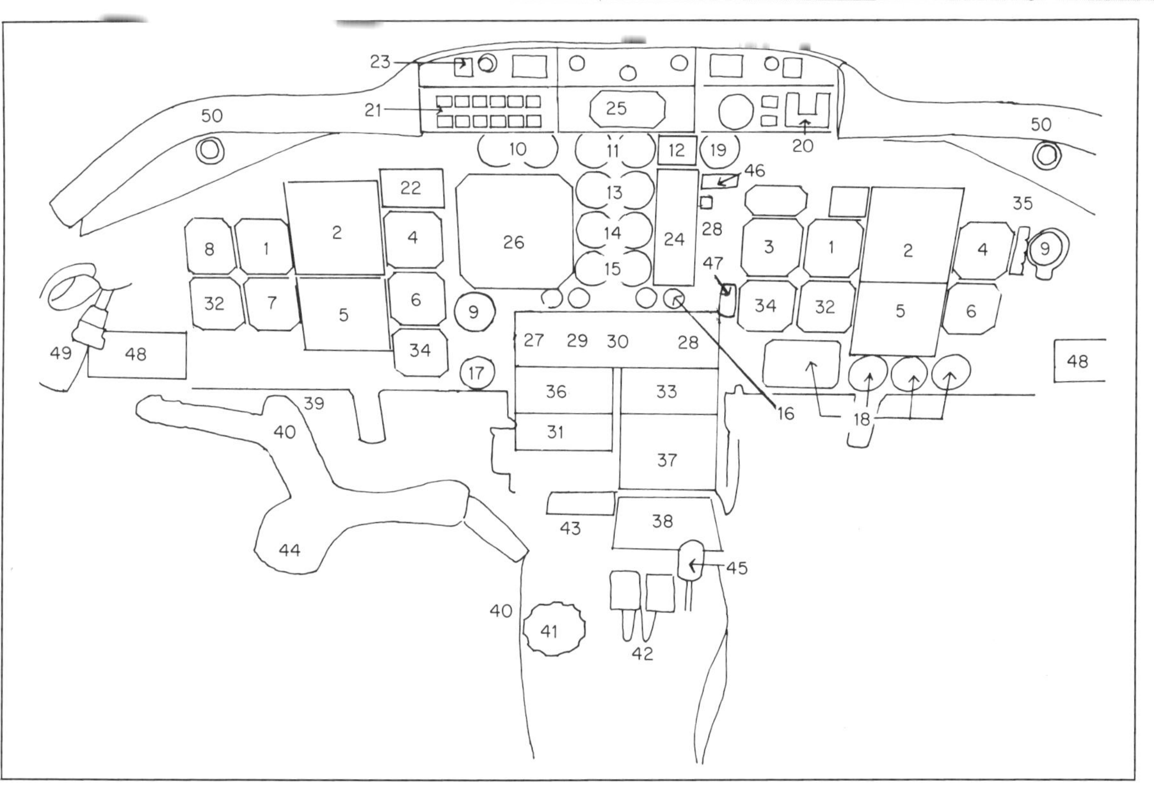
23
21
25
50
50
20
10
11
12
19
46
22
13
35
8
1
2
4
26
14
24
28
3
1
2
4
9
15
47
32
7
5
6
9
34
32
5
6
27
29
30
28
49
48
34
17
48
36
33
39
16
18
40
31
37
43
38
44
45
40
41
42

Figure 40 HS 125 700 Series, world-beating business jet built by British Aerospace at their Chester factory. *British Aerospace*

What are instruments for anyway? Hang-gliders, for example, don't have any. You will realise if ever you get your hands on the controls that it is fairly easy to fly without instruments when you can see where you are going. You can navigate by means of landmarks and use the horizon to keep straight and level. Once you get into cloud, though, everything is suddenly very different.

The horizon is invisible, so you cannot tell whether your wings are level or whether the nose is pointing up or down. You cannot see whether you are keeping to a straight course or going round in circles. So in order to control your attitude you need the two very important instruments in the middle of the Basic T – the *Artificial Horizon* and the *Compass* or its equivalent.

The artificial horizon is an instrument consisting of a circular card driven by a gyroscope. The lower half of the circular card is black and the top is pale-coloured. Whatever becomes of the aircraft's attitude – it can bank or dip – the gyro will hold the card stable and you can fly on it just as you would on the real horizon. The artificial horizon, or some more sophisticated attitude indicator based on it, is the most important instrument for blind-flying, and for this reason occupies the centre of the Basic T. Just below it is the Compass, or one of its derivatives: Direction Indicator, Horizontal Situation Indicator, or whatever. Now you can not only fly level, but straight as well.

Even if we had perfectly clear skies, we would still need instruments to monitor the aeroplane's performance. The two most important are the *Altimeter* and the *Air Speed Indicator*, usually placed on the right and left arms of the T respectively. Each of these instruments deserves a section to itself later on. Brief mention must be made of the few remaining flight instruments which are found in most aircraft. One will be the Vertical Speed Indicator, which measures the rate-of-climb (or descent), another will be the Turn-and-Bank Indicator. This enables the pilot to balance out his turns without slipping sideways into the turn (caused by too much bank) or skidding to the outside of the turn (caused by too much rudder). In its simplest form it consists of a small spirit level. A useful additional instrument is a Rad Alt (short for Radar – or Radio – Altimeter). This measures height (not altitude) by bouncing radar pulses off the ground and timing their return.

Dual controls are found in many aircraft, but not always full dual instrumentation. The Captain (or whoever is flying the plane – not always the same person) is expected to sit in the left-hand seat, where there is a full set of instruments, or in the front cockpit of a tandem layout, such as in the RAF's Hawk trainer. If the pilot is a student or trainee, the instructor occupies the less-well-equipped right-hand, or rear, station.

Flying Circuits

Training flights often consist of circuits where the pilot wishes to get in as much practice as possible at approaches, landings, take-offs, and low-speed and low-level handling generally. First the take-off followed by a climb to circuit height, which is rarely more than 1000 feet. Circuits are rectangular, and each leg has a special name; after take-off and climb a turn will be made onto the *Crosswind Leg*, and the climb continued if necessary. Next follows the *Downwind Leg*, where as a standard drill the pre-landing checks are carried out. Next the *Base Leg*, followed by the *Final Approach*. At some airfields circuits are always flown the same way round. After practising an approach, the pilot may pile on the power and go round again – an *overshoot* – or he may touch down first and then rev up – 'Touch and Go'.

Jargon borrowed from circuit flying can be used to describe any approach and landing – thus we hear of Left Base or Right Base depending on which way one turns onto final approach (Turning Finals, this is called). The final approach is called Finals and a Finals may be either Short or Long.

Wind

Flying would be a lot easier if there were no wind. It complicates take-offs and landings, especially if it is a crosswind, or if it is gusting. Once airborne you are blown along by it, more surely even than a sailing yacht on the open sea.

Of course you rarely notice this. Most aircraft have a performance adequate enough to fly into the teeth of a gale and still make good time to their destination. If your airspeed is 115 knots and the headwind is 35 knots against you, then you are still making good 80 knots over the ground. On the way back you will have the wind behind you and race home at 150 knots!

A strong crosswind blowing across your track will try to force you off course, so you counteract this by steering into it. The angle you have to steer into a crosswind is termed *Drift*, and from the ground it looks as though an aircraft is flying sideways. Watch light planes flying in a strong crosswind or large aircraft landing.

Cloud and Visibility

Although take-offs and landings may be affected by low cloud, fog or haze, once airborne, cloud itself causes few problems, as most aircraft carry blind-flying instruments. Visibility is measured along the ground in metres (or kilometres), the height of the lowest cloud layer (the cloudbase) is given in feet, and the amount of cloud in *octas* or eighths of sky covered. As mentioned, if the runway visibility (RVR – Runway Visual Range) is below 600 metres, many airports close. If the cloudbase is very low this may limit pilots who are not qualified (or equipped) to make an instrument landing. Thus continuous weather bulletins are broadcast on a special VHF channel called VOLMET, giving conditions at airfields round the country. In bad weather pilots tune to VOLMET continually, trying to decide whether to press on or to divert to another airfield.

As we have explained, keeping straight and level in cloud is no problem. Using radio aids, navigation is fairly easy, even when there are no landmarks visible. The real problem is air safety. There you are bowling along in the clouds without a care in the world; how do you know someone isn't going to crash into you?

In good flying weather you can see for miles around and you can also see the ground below. This is called *VMC – Visual Meteorological Conditions* (Victor Mike Charlie). You can see and be seen, therefore you can look out for yourself – in theory at least. You keep clear of other aircraft by applying *VFR – Visual Flight Rules*. In cloud, which pilots call *IMC – Instrument Meteorological Conditions*, or India Mike Charlie, you use IFR, that is *Instrument Flight Rules* instead. These mainly state that aircraft flying in different directions should cruise at different levels (see Chapter Six). In busy airspace it is necessary to use the services of Air Traffic Control, and sometimes this applies even if conditions are VMC. Controllers using radar keep aircraft separate, and pilots are obliged to obey their instructions. The operating rules for large commercial aircraft usually specify that IFR and the services of Air Traffic Control shall be used at all times.

Weather

There are only three kinds of weather which need worry pilots (in the air at any rate – what happens on the runway is a different matter). The first of these, ice, is rarely a problem as today's planes are usually equipped with de-icing devices. Without de-icers a pilot would need to avoid icing-up by flying low enough for the ice to melt, or high enough, where there is too little moisture to freeze into ice, the air being cold but dry.

The second and third kinds of weather, storms and turbulence, often go together. In broad daylight the storms show up as great heaped-up masses of cloud, often rising higher than the plane can climb. This kind of cloud is called *Cumulo-Nimbus* and well-developed examples spread out on top into huge anvil-shaped heads. The body of the cloud is an evil greyish-purple mountain, inside of which all hell is being let loose. A pilot who flies through one of these is in for a rough ride indeed. They contain rain, hail, lightning and turbulence. In severe storms the hailstones are big enough to damage the aircraft, and the turbulence is rough enough to stand it on its head.

Pilots can steer clear of cumulo-nimbus if they carry *Weather Radar*, because these clouds have a characteristic appearance on the radar screen, and the more severe the storm, the more it shows. At one time it was considered an annoying feature of short-wavelength radar that it could pick out clouds, rain and hail, but carried in an aircraft and used for spotting thunderstorms, this disadvantage has been turned to good purpose. The beam of the scanner sweeps ahead of the aircraft in an arc 30° to either side. Directed onto the ground it may sometimes produce a recognisable map of the terrain below, and can thus be used for navigation also!

Unfortunately turbulence is not always found in recognisable cloud formations. *CAT – Clear Air Turbulence* – is a feature which pilots often report. You cannot see the extent of it, so you carry on flying through it. The plane should take it without any trouble but a captain's main worry may be that one of the passengers has not fastened his seatbelt really tight – there have been cases of serious injury in CAT.

Measurement in Aviation

Before we go on to discuss aircraft performance we need to note how it is measured, and in what units.

Flying has borrowed many of its systems from seafaring and these are still used. Distance is usually measured in Nautical Miles, rather than ordinary Statute Miles, speed is very often quoted in Knots (nautical miles per hour) and position is described in terms of latitude and longitude.

Position Latitude and longitude can be used to refer to any position on the globe, land or sea, or in the air above it. The system is based on angular measurements taken from the centre of the globe, and the unit of measurement is the *Degree*. Each degree is subdivided into 60 *Minutes*, and each Minute into 60 *Seconds*. On some charts an alternative system is used of subdividing the minutes decimally, and abolishing seconds altogether. Thus the position of the VOR/DME beacon near Daventry may be given as:

either 52°10′46″ N 01°06′44″ W
or 521046N 0010644 W
or even 52°10.8′ N 001°06.7′W

The system looks more complicated than it really is. With a bit of practice at plotting, say, the positions of radio beacons, you will soon become familiar with it.

Lines of latitude are known as *parallels* and lines of longitude as *meridians*. The lattice which these lines make on a globe or chart is known as a *graticule*. If you look at a globe you will realise that a degree of latitude is the same length anywhere on the Earth's surface: 69.1 statute miles. However, the length of a degree of longitude is this figure multiplied by the cosine of the latitude at that place.

The Nautical Mile is the length of a minute of latitude, which works out at 6080 feet, 1.152 statute miles, or 1.853 kilometres. In aviation, the nautical mile is used to measure navigational distances. Pilots when speaking of distances are almost always referring to nautical miles, not the statute miles familiar to the motorist.

The following conversions are useful:

1 nautical mile (n.m.)	= 6080 feet
	= 1.152 statute miles
	= 1.853 kilometres
	= 1 minute of latitude
1 degree of latitude	= 60 minutes of latitude
	= 60 nautical miles
	= 69.1 statute miles
	= 111.2 kilometres

The length of a degree of longitude varies depending on the latitude. For any given place look up the *cosine* of the angle of latitude in a set of mathematical tables and multiply this by 60 n.m. or 69.1 statute miles to find the length of a degree of longitude at that place. A typical value for the UK (e.g. Manchester) is about 36 nautical miles.

Speed Speed is usually quoted in *knots*, which are nautical-miles-per-hour. Knots are the units preferred by Air Traffic Control for giving wind and aircraft speeds. In some places miles-per-hour are used; to convert knots to m.p.h. multiply by 1.152. In an aircraft the 'speedometer' is the *Airspeed Indicator* (ASI), which measures the speed at which the aircraft is moving through the air. It does this by measuring the air pressure in a small tube called the *pitot head* which points forward into the airstream. The method is reliable, but not particularly exact. The figure shown on the ASI is what is termed the *Indicated Air Speed* (*IAS*). To obtain the *True Air Speed* (*TAS*), certain corrections for outside air temperature must be made, usually by setting a small sub-scale on the instrument itself. Groundspeed, the speed at which the plane is covering distance, depends on the wind component and cannot be measured directly unless electronic devices such as radar, Doppler or DME are used.

Mach Number A Mach number is airspeed relative to the speed of sound under comparable conditions. High performance aircraft need to think of speed in terms of Mach numbers, because as the speed of sound is approached and passed, shock waves build up which tend to stay with the aircraft, causing dangerous buffeting and a tendency for the controls to misbehave. In order to keep an eye on the Mach number a special indicator called a *Machmeter* is fitted to almost all jet aircraft. Some arrangements combine the Machmeter and ASI in a single *Combined Speed Indicator*.

To obtain the Mach number, divide the airspeed by the speed of sound, but this varies depending on the temperature. Under ISA conditions the speed of sound at sea level is 762 m.p.h. but in the much colder air of 35,000 feet where many aircraft fly the speed of sound (Mach One) drops to 660 m.p.h. (573 knots). Thus an aircraft cruising at this altitude at 600 m.p.h. should be registering something like Mach 0.91. Concorde's cruising speed of 1350 m.p.h. works out at just over Mach 2, twice the speed of sound.

Speed Conversion Table

Knots	*m.p.h.*	*Nautical miles per minute*	*Seconds per nautical mile*	*Mach number*	
				Sea level	*35,000'*
100	115	1.7	36		
150	173	2.5	24		
200	230	3.3	18		
250	288	4.2	14		
300	345	5.0	12		
350	403	5.8	10		
400	461	6.7	9	0.61	0.69
450	518	7.5	8	0.68	0.78
500	576	8.3	7	0.76	0.87
550	633	9.2	$6\frac{1}{2}$	0.83	0.96
600	690	10.0	6	0.9	1.05
700	806	11.7	5	1.06	1.22
800	921	13.3	$4\frac{1}{2}$	1.21	1.40
900	1036	15.0	4	1.36	1.57
1000	1152	16.7	3.6	1.52	1.75
1200	1382	20.0	3	1.82	2.09

Performance Figures

Do aircraft fly as high or as fast in practice as they say they can? People who may marvel at the performance figures quoted for modern aircraft are even more surprised to learn that in fact it pays to fly high and fast. An observer studying a contrail in the clear blue sky seven or eight miles overhead notices the silvery unreal wraith of an aircraft at the head of the column. It may look incredible: 350 tons of Boeing creaming along at 500 knots at 35,000 feet, but to the airlines it is neither unreal nor incredible, just everyday economics. Jet engines burn less fuel per mile at these altitudes, and speed pays dividends too.

As petroleum supplies look like running out eventually, and as even today an airline's fuel bill is a major expense item, methods of saving fuel become more and more important. Some companies run schemes to encourage their pilots to save fuel. They could do this by cutting back on speed. On short-haul flights the passengers probably don't notice, and it is known that some European airlines knock about 50 knots off the cruising speed on short flights within Europe. The difference isn't noticeable when set alongside the usual take-off delays, landing delays, customs delays, etc., which are notorious on some of these routes.

On long flights, however, the passengers would notice any cut-back in speed. Their wristwatches would soon tell them that an extra hour had been added to an otherwise 14-hour flight, and next time they might book with another airline, because after that length of time trying to sleep in a cramped seat their one wish is to get down again to relax in more spacious accommodation. Long-distance flights therefore try to stick to their published schedule. One way to do this is to pick up a strong tailwind, or at least to avoid headwinds. In some places the 'jet-stream' winds can be blowing as strongly as 150 m.p.h., even

at the altitudes at which airliners fly.

The principle 'the higher the better' operates all along the line. Even piston-engined aircraft which had a ceiling around 25,000 feet found it paid to climb as near to this as possible on long flights. With jet-engines, which reign supreme above 400 knots and above 30,000 feet, the trend continued. Jet aircraft of all types are at their economical best between 35,000 feet and 40,000 feet. Concorde, cruising at 50,000 feet, goes even higher. In fact, so much does altitude improve performance that some short-haul airliners, such as the Boeing 737, prefer to get up to their best and most economical cruising altitude even if it means spending all the first half of a short flight climbing and all the second half descending, a flight profile which is not too popular at all with air traffic control, because it means that all the intervening flight levels must be kept clear.

Accurate performance figures for many military aircraft are not readily available; they count as classified information in some cases, but the situation in practice is modified by a number of factors. The first of these is the tendency to operate many aircraft with additional fuel tanks, missiles, bombs, etc., slung underneath the wings and fuselage. This is bound to detract from the optimum 'advertised' performance figures which may apply to the aircraft type only in the 'clean' configuration.

Also the luxury of flying at the fuel-saving altitudes of 40,000 feet must be denied to many military planes nowadays, whatever their role. In order to get in under enemy radar without being spotted it is necessary to fly low, hugging the landscape. An entire mission may need to be flown in this way, and the need to fly low applies not only to strike aircraft such as the Jaguar, but to support aircraft such as the Hercules, and even to interceptors, which must train on the assumption that their quarry would attack at low-level. Not only is low-level training vitally necessary, but it uses more fuel.

Altimeters

There are two kinds of altimeters. The *Rad Alt* (Radio or Radar Altimeter) works by bouncing radio signals off the terrain and gives a continuous indication of the aircraft's height above the ground. It reads height, not altitude, and is particularly useful when approaching to land over hilly country or for certain military low-

Figure 41 Contrails (or Vapour Trails) are formed by the passage of an aircraft through super-cooled air. A large aircraft such as this will also leave a turbulent wake or vortex which is best avoided by other aircraft. *British Airways*

flying purposes. However, not all aircraft are equipped with Rad Alts. Much more common, and in fact almost universal, is the instrument known as a *Sensitive Altimeter* which works by measuring air pressure. As is well-known, air pressure decreases as one ascends, and a modern altimeter will register the pressure difference between the floor and the top of a table.

Inside the instrument is a metal box which is sealed and contains a partial vacuum. The varying pressure around this box forces the sides in and out against a spring, and the small movement of the box is amplified by an arrangement of levers and clockwork, driving the several pointers on the face of the instrument. In a *Servo Altimeter* the linkage is electromechanical and the altitude readout can be shown digitally.

Normal air pressure at sea level is somewhere around 1013 millibars. This decreases by 1 millibar for every 30 feet we ascend. However, if you study the newspaper or television weather charts you will see that the pressure at sea level could be as high as 1040 mb during a heatwave, or as low as 960 mb during a storm. This variation is not only from day-to-day but from place to place. A pilot flying from A to B could misread his altitude by as much as 30 times 80, that is 2400 feet. To overcome this difficulty there is provision for all pressure altimeters to be set or zeroed according to the sea-level air pressure at any place. The *setting knob* fitted to the instrument can be used by the pilot to set the correct pressure on the *millibar sub scale* so that the altimeter will show the correct altitude. There are a number of different disciplines for setting altimeters, such as QNH, QFE and ISA.

The pressure (in millibars) which is set up by the pilot on the altimeter sub-scale may be either:

(a) The actual pressure on the airfield or at the runway threshold. An altimeter set up with this pressure will read zero just as the plane lands. Such a setting is called a *QFE*.

(b) The pressure corrected for mean sea level. Such a setting is called a *QNH* and will cause the instrument to read true *altitude*, that is, height above mean sea level.

(c) A standard pressure setting of 1013.2 millibars. This is based on the International Standard Atmosphere (ISA).

Both QNH and QFE are what are termed 'Q-code' phrases, dating back to the early days of wireless. A pilot tapping out the letters QNH or QFE in Morse would be given the correct altimeter setting by the controller in reply. Nowadays each codeword means not the query, but the reply, and each is a precise technical term in its own right.

QFE An altimeter set to this figure will show the height above the airfield – very useful when approaching to land.

QNH An altimeter set to this figure will show the correct *altitude*, that is the height *Above Mean Sea Level* (*a.m.s.l.*). The mean sea-level datum is the average of low spring tides in Portland Bay – in case you need to know! – and the heights shown on maps, charts, etc., are related to this datum. Thus an altimeter set to the QNH is necessary for flying across country. UK airspace is divided into 19 *Altimeter Setting Regions*, for each of which a different QNH is promulgated. Each of these Regional QNH pressure settings is valid for an hour, and is a millibar or two lower than the lowest expected value. The Altimeter Setting Regions (*ASRs*) are shown on the sketch map. When an aircraft is preparing to land, the pilot is given a *Local QNH* by the controller. In practice he is given both QNH and QFE settings and is supposed to read them back over the R/T. If he has two altimeters he sets the QNH on one and the QFE on the other. One altimeter will show his height above the airfield and the other will show his height above sea level.

The millibar readings given in weather forecasts are corrected for mean sea level and should be the same as the QNH.

Finally there is the ISA pressure setting of 1013.2 millibars. Above 3000 – 4000 feet in the UK *all* aircraft set their altimeters to this figure. There is little risk of them colliding with the terrain, but every risk of them colliding with each other unless they fly with synchronised altimeters. The 500 or 1000 foot minimum vertical separation which Air Traffic Control tries to arrange would be meaningless if all altimeters were set differently.

An altimeter set to 1013.2 millibars ISA above 4000 feet will not always indicate true altitude; the reading is known instead as a *Flight Level*, and is rounded to hundreds of feet. Thus a reading of 6025 feet becomes FL 60, or on radio 'Flight Level Six Zero'. Similarly FL 330 is an Air Traffic Control way of saying 33,000 feet.

Direction

Direction in a horizontal plane, or in *azimuth* as it is called, is measured in degrees of arc, working clockwise around a 360° compass rose. The old-style mariner's Compass Rose with its Nor'Nor East and all that, has been abandoned long ago. So now we have a

Figure 42 Altimeter setting regions of the UK.
Reproduced by courtesy of RAF No.1 AIDU

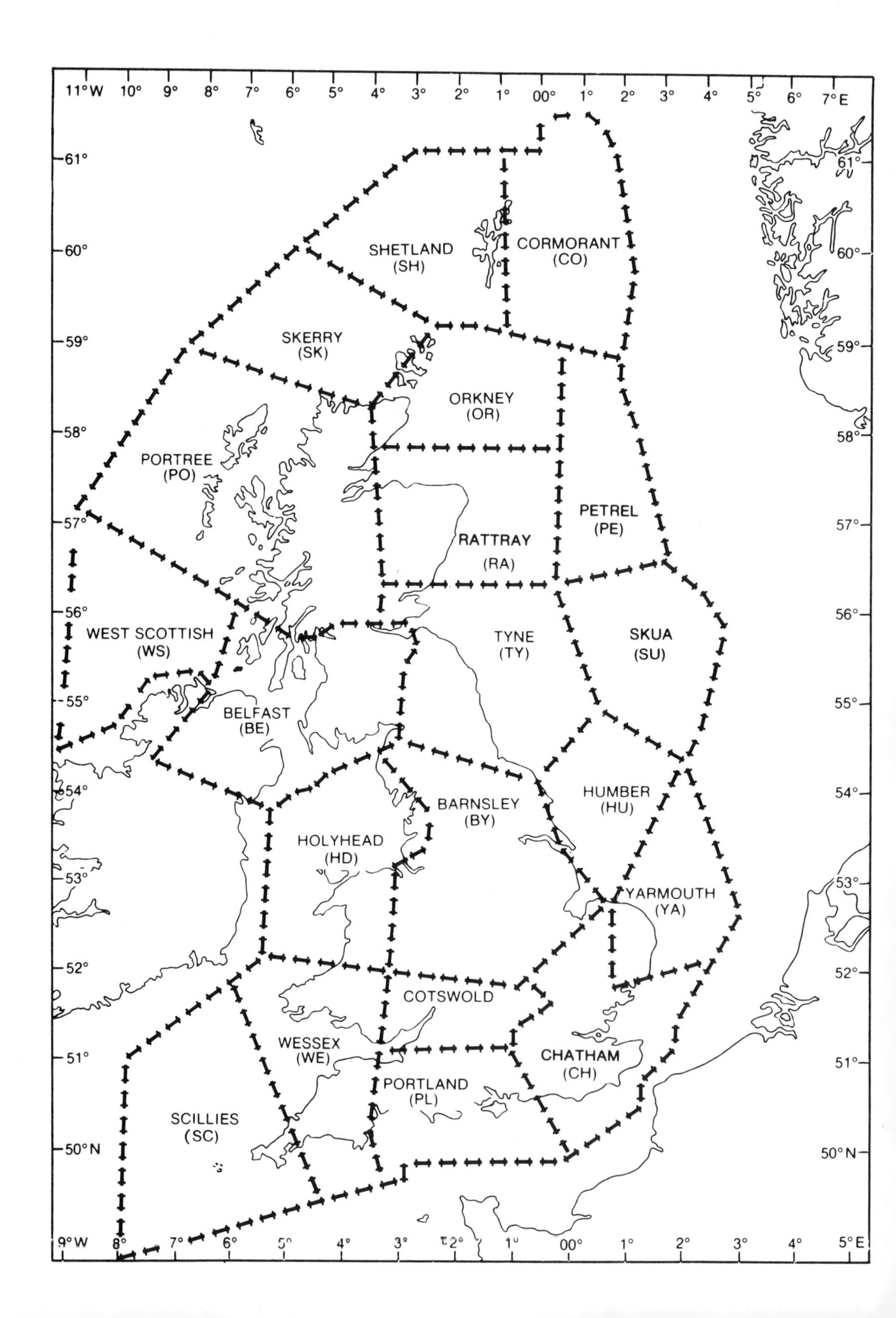

neat, precise, numerical method of giving courses and bearings. North is both 0° and 360°, east is 090° and southwest is 225°. The direction in which an aircraft's nose is pointing is its *heading*. Its course over the ground is its *track* and the direction of any station or object in relation to the aircraft is a *bearing*. Measured from the other end, we get a *back-bearing* or *reciprocal*; just add (or subtract) 180°.

Direction is measured in relation to magnetic north. There are other methods of wayfinding, but it must be humbling to think that large modern aircraft still fall back on magnetic compasses, like mediaeval explorers or present-day Scouts. Headings, bearings, etc., are invariably given in relation to magnetic north rather than true north.

'Echo Delta. Steer one-five-zero,' the controller instructs crisply. The pilot turns the plane and the card of the magnetic compass rotates until 150° is squared off underneath the lubber-line. Even the bearings given by automatic radio beacons such as VORs are in magnetic degrees rather than true. As magnetic north moves around, so these beacons have to be re-calibrated.

At the present time, over most of the British Isles, the *Magnetic Variation* is about 8°W, which means that magnetic north is about 8° west of True North. Lines on a chart showing the magnetic variation at any place are known as *Isogonals*.

QDM and QTE These are more phrases from the Q-code, popularised by pilots who needed help from a radio direction finder on the ground when they had got themselves lost! *QDM* means 'Please give me a magnetic track I can steer to reach you.' A QDM is measured from the aircraft, and is magnetic. A *QTE* is measured from the ground station to the aircraft and is a true bearing, not magnetic.

Compasses and Direction Indicators

An ordinary simple magnetic compass is too susceptible to the presence of ferrous metals and electrical systems for it to be fitted on an instrument panel within the Basic T. The problem is solved in several ways, depending on how sophisticated the aircraft is:

1. A compass is mounted in the centre of the windscreen.
2. As above, but a gyrocompass (*Directional Gyro Indicator, DGI or DI*) is mounted in its rightful place in the Basic T. This has to be set by the pilot to match the magnetic compass.
3. A servo-compass is fitted in the T. With this type the magnetic sensors are out on the wingtips, clear of most interference. A gyro input to the compass is an added luxury which improves its performance.
4. A *Horizontal Situation Indicator* (*HSI*) is not only a sophisticated magnetic-cum-gyro-compass, it also takes inputs from electronic course-steering and landing aids, i.e. VOR and ILS. Thus the pilot has in front of him directly within his gaze and just below the AH, all the information he needs to steer a course.

Automatic Flight

95% or more of the time flying is a routine task consisting of little more than keeping the aircraft straight and level. Autopilots worked by gyroscopes can do as much. They are sensitive to any change in the aircraft's heading or attitude and can correct it by applying the controls. Having set the autopilot the human pilot's hands are free to attend to other tasks.

Modern autopilots can be set to lock onto a number of different readings or features. Besides the compass (heading) lock, the autopilot can be programmed to lock speed, attitude, altitude, ILS, VOR, ADF, etc. It may control engine speed by using *autothrottles*, and it may be happy to take orders from a navigation system such as INS, Omega or R-Nav. However, the human pilot can take over at any time. Built into the handle of his control column is the Autopilot Disconnect button. One blip on this and all the locks drop out; he has the plane back in his hands again.

But flying straight and level isn't everything an autopilot can do. It can compute and execute course changes, such as Rate One turns to intercept a new VOR radial or an ILS beam, and manoeuvre the aircraft so smoothly that the passengers don't even get to know it. It can do this with the autopilot itself fully in control (*Flight Control System*) or it can supply guidance for the pilot to operate the controls (*Flight Director System*). With both arrangements the autopilot needs to be wired into the main attitude instruments within the Basic T – The *Attitude Director Indicator* (*ADI*) and the *Horizontal Situation Indicator* (*HSI*).

A Flight Director System guides the pilot by means of a system of crosses, Vee-bars, flying wings etc., within the ADI. For example, if the cross moves, the pilot chases it with the controls until he has brought it back! In the HSI a deflection pointer may move to left or right, indicating a turn either way. Flick a switch perhaps and the autopilot takes over, resulting in a full Flight Control system. The human pilot may still monitor the instruments, but he has his hands free to eat his lunch, keeping the Autopilot Disconnect button clearly in view.

Figure 43 Heading and bearings

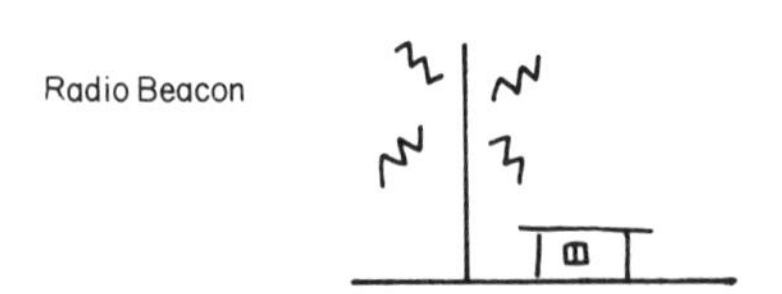

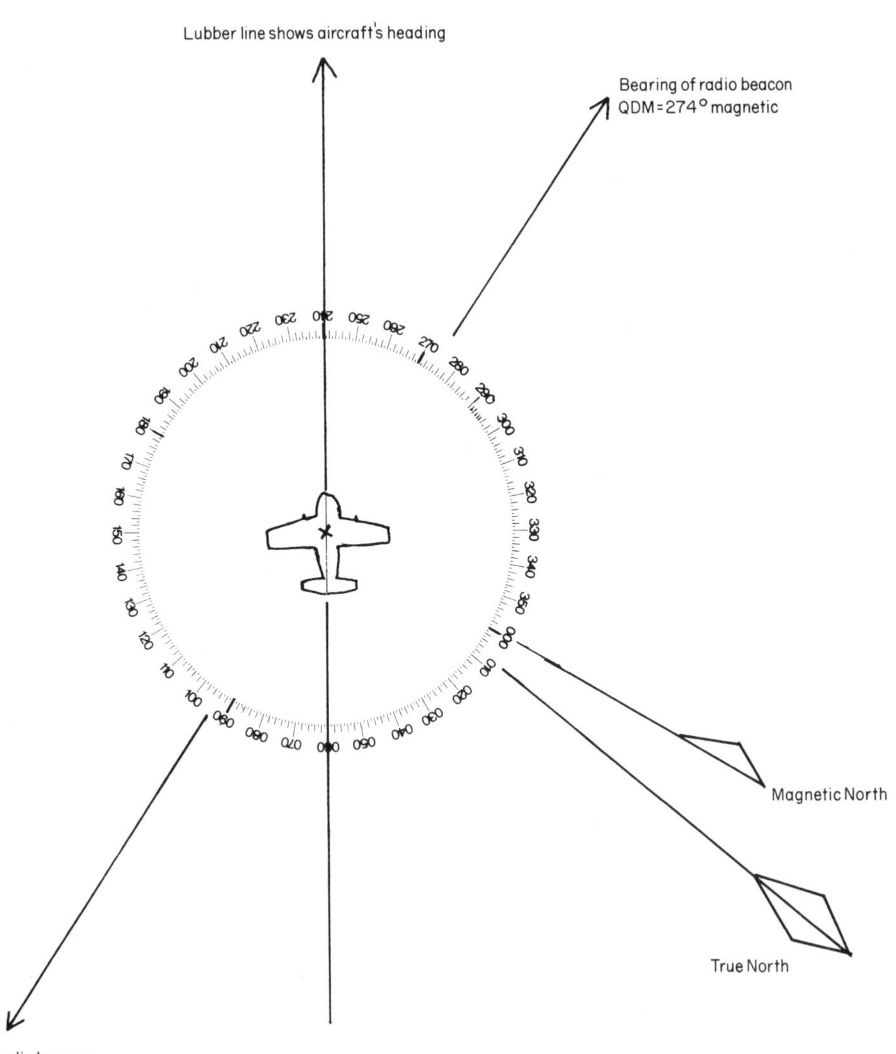

Basic Aerial Navigation

You can do this for yourself even if you are only a passenger. You need a suitable map or chart and you need to know the exact track the aircraft will fly. Draw a thin pencil *Rhumb Line* to represent your track from A to B. As you fly along this you can note the landmarks. Those which show up best from the air are the conspicuous ones: coastal features, lakes and reservoirs, rivers and canals, motorways, railways, power stations, disused airfields, etc. You use these landmarks to plot your progess, to fix your position and to ensure that you are not wandering to left or right of your track.

You can make it more complicated by clocking your time off against distance and airspeed. This will enable you to check groundspeed and possibly the wind. If you know how to use a wind computer you can tell the pilot the course to steer to maintain his track in a crosswind. Get all your sums right and he might ask you to fly with him again!

Topographical Charts

The ideal map to use for the above exercise would be something like a Quarter Inch to the Mile Ordnance Survey Map – a scale of 1:250,000. This would show roads, railways and other features which could be observed from the air. A special edition of these maps overprinted with a graticule, showing airfields and other details of aeronautical significance, is published by the Civil Aviation Authority as a series of 18 Topographical Air Charts of the UK.

These charts are extremely beautiful publications, and worth acquiring and collecting for yourself, even if you have to buy them new. They may be obtained from your local airfield or by post from a specialist supplier such as Edward Stanford, 12–14 Long Acre, London WC2. Some enthusiasts even use these charts for motoring!

Another series of charts is published by the CAA also, to a scale of 1:500,000 (half-million). These charts do not show as much topographical detail, but there is room on them to show beacons and other details of radio navigation. A set of four 'half-mill' charts cover the UK. For a current price list of CAA charts, write to the Aeronautical Information Service, Tolcarne Drive, Pinner, Middlesex HA5 2DU.

Flight Guides

Apart from a set of charts a well-equipped pilot needs a booklet of the kind known as a Flight Guide. This gives a wealth of important aeronautical information, especially details of airfields. The Daddy of all Flight Guides is the *UK Air Pilot*, a huge heavy looseleaf binder published by the CAA, as is its offspring, the *General Aviation Flight Guide*. The *Aerad* publishing house, now run by British Airways, publish their own flight guide and bulletin to supplement their radio navigation charts. Light plane pilots find *Pooley's Flight Guide* useful as it lists many small airfields. However, for the enthusiast the most economical and compact booklet is likely to be the *En-Route Supplement* (*Northern Europe and North Atlantic*) which is published by the RAF at £3. Write to No. 1 AIDU, RAF Northolt, West End Road, Ruislip HA4 6NG. Even if you are not a pilot, just a mere enthusiast, these little booklets are a mine of information.

Dead Reckoning

This is a method of navigating blind by working out your track, time and speed from your last known position. The big problem is wind. If you don't know its speed and direction you will not be able to make accurate allowance for crosswind drift, hence you won't be sure of your exact track. These corrections are made by using a little plastic gadget for solving vector triangles called a wind computer or *navigation computer* and costing only a few pence. Also, your groundspeed along your track will depend on the wind component. That little computer again. Using Dead Reckoning (DR) as it is called, you work out your position as you go along, all the time hoping that you are where you think you are.

Suddenly through a gap in the cloud you see a familiar landmark. Now you *know* where you are. This is called a *fix*, and you can use it to update your dead reckoning. Not only can you start all over again from this fix, but you can backtrack on your earlier calculations to check if you got the wind right! Dead reckoning is hard work. Most pilots cheat by using frequent fixes plotted from radio beacons.

However, a military pilot cannot expect to rely on radio fixes, so he often has an electronic computer to do his DR for him. These are of two kinds. An *Air Position Indicator* keeps up a DR plot by taking inputs from the ASI and the compass. The pilot still has to correct for wind. A *Ground Position Indicator* on the other hand takes inputs from Doppler radar as well as from the compass, and therefore corrects automatically for wind drift. For details of Doppler, see Chapter Five.

Celestial Navigation

This is a method of position-fixing by using a sextant to measure the angle of elevation of the sun

Figure 44 Topographical aircharts of the UK

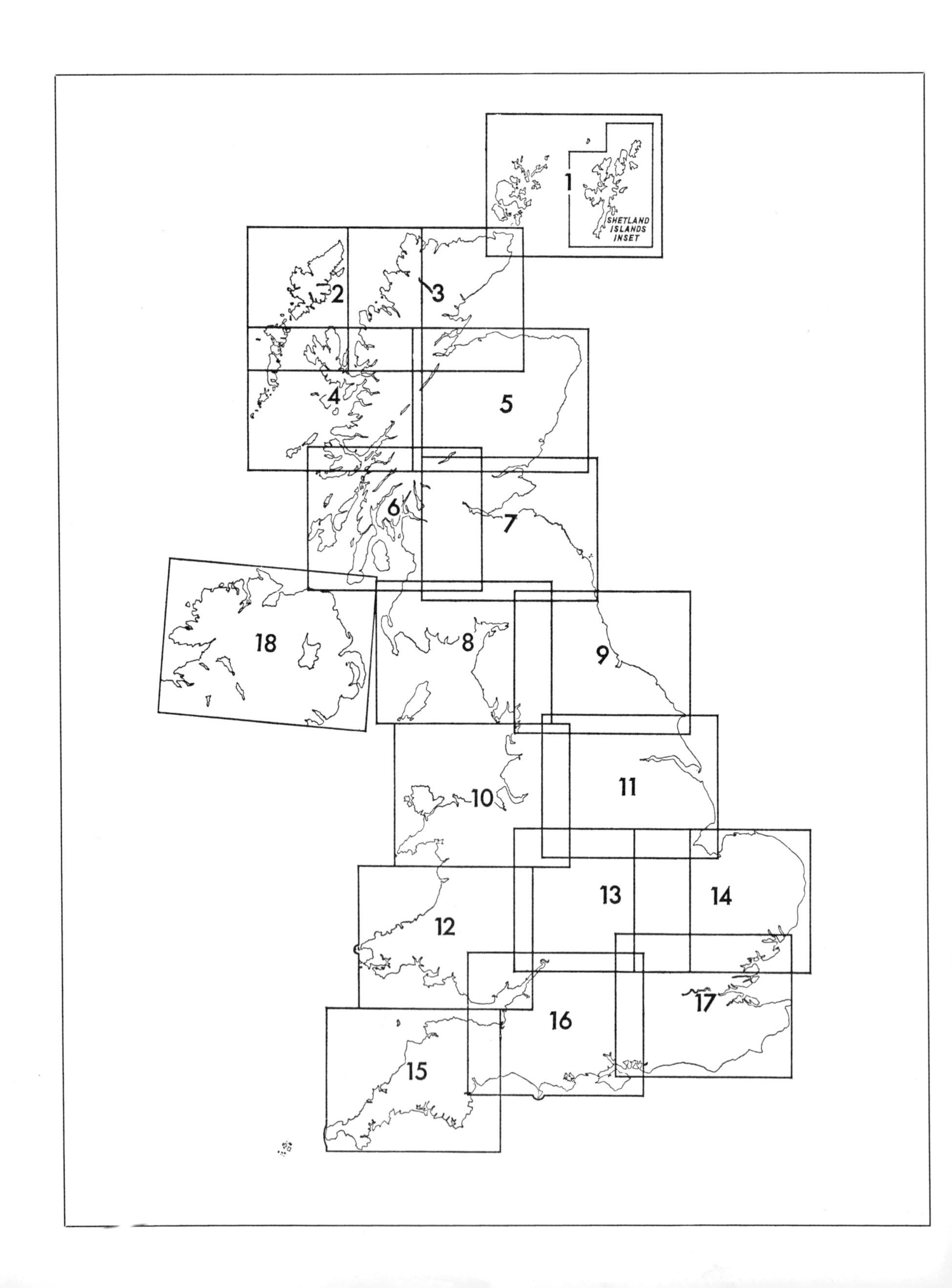

Figure 45 A Puma helicopter on North Sea oil business lands in the Nordraug platform. *A Shell Photograph*

and stars. It is a complex and exacting discipline, and requires moreover that the aircraft should have a transparent *astrodome* fitted atop the fuselage. In a pressure hull an astrodome would be a weak point, so a pressurised mounting for a periscopic sextant would be fitted instead.

Helicopters

Although most of the preceding section has been written with fixed-wing aircraft mainly in mind, much of what has been said applies also to helicopters. Although helicopters do not fly as high or as fast as their fixed-wing counterparts, they score by being able to fly safely at much lower altitudes. They can land and take off almost anywhere and can pull off daring rescues from tight situations. They have proved their value in warfare in aggressive as well as in support roles.

Helicopter flights across country are usually made well below 1000 feet. Following the terrain is usually easy. But when the weather closes in, a chopper is unable to proceed; some of the instrument flying and navigational devices perfected for fixed-wing aircraft do not work quite so reliably on helicopters. Doppler, however, works quite well and has been developed into a reliable fully automatic hovering device.

Figure 46 A Sikorsky S61A helicopter. *British Airways*
Figure 47 A Wessex helicopter delivers supplies to HMS Fearless. *RN*

Things to Do

1 Try to find an aircraft instrument panel which you can inspect, and then look for the following:
  (a) Control column
  (b) Rudder pedals
  (c) Engine and propeller controls
  (d) Basic T flying instruments: ASI, Artificial Horizon (or ADI)
  Altimeter and Compass (or DI or HSI)
  (e) Other flying instruments
  (f) Radio and navigation instruments.

2 Observe aircraft in flight as follows:
  (a) Training aircraft flying circuits
  (b) The effect of wind on aircraft, particularly on the landing approach; Which way does an aircraft steer to counteract a crosswind? Is turbulence more obvious near the ground?
  (c) In suitable weather look for vapour trails. Do they always follow more or less the same tracks, indicating the presence of an airway or a military route?

3 Try to find the latitude and longitude of your home or of a local airfield.

4 Try to obtain the sheet of the UK 1:250,000 series of Topographical Air Charts which covers your area and study it. Note the locations of Control Zones, MATZ, Danger and Restricted Areas, etc. Why are these areas dangerous?

5 Collect QNH values (air pressure at mean sea level) from newspaper and television weather maps, or from an aircraft radio, and note how they are high in fine weather and low in bad weather.

4-Aircraft Radio

Although the Wright Brothers invented the aeroplane, and Sir Frank Whittle is among those who can claim the credit for having produced the first jet engine, aviation as we know it today would be impossible without Marconi. Radio in today's aircraft is indispensable not only for communication but as the basis of navigation and landing systems. Thanks to transistorisation and miniaturisation of circuits and components, even single-engined light aircraft can be fitted with radio communication and navigation equipment to an extent which would have been unforeseeable even thirty years ago. As solid-state circuitry has taken over from valves and switches, equipment has not only shrunk in size and weight, it has also become more reliable and easier to use. Nowadays the game is proceeding one stage further as more and more of the miniature computing circuits called microprocessors are beginning to be employed, particularly in navigation systems.

The main purpose of this short chapter is to explain the intricacies of aircraft communications radio, or R/T as it is often called. In passing we will need to touch on some of the properties of radio waves and how they are transmitted and received.

The Radio Panel

An aircraft's radio equipment is usually located in the centre of the instrument panel where it can be reached by either pilot. There is no standard complement or layout. For airways flying the following minimum would be the rule:

2 radiotelephone transceivers (R/T) for voice communication
2 VHF NAV receivers for navigation and landing aids
1 Automatic Direction Finder (ADF) with its own tuner
1 75 MHz marker beacon receiver
1 DME set (Distance Measuring Equipment)
1 ATC transponder for identification on radar

Each of these communicates with the outside world, so it has to have an aerial (or antenna, as the Americans call it). You will find that the aerials are scattered all over the outside of the aircraft, some short, some long, while others are fitted flush beneath dielectric plastic panels.

Radio Waves

Radio waves are electromagnetic waves in space which are produced by an alternating voltage in a suitable circuit and fed to an aerial. These waves travel outwards at the speed of light and if another circuit, tuned to the same carrier-wave *frequency* is placed in their path, this receiver will resonate with the transmitter and reproduce faithfully any modulation pattern which was applied to the carrier wave at the transmitter. This modulation can be in the form of speech or music, navigational data or even a television picture.

The carrier wave remains steady, alternating (or oscillating) from a positive to a negative voltage and back again a given number of times per second; this is its frequency, measured in Hertz (Hz). We usually speak of thousands of Hertz, or kilohertz (kHz), or even millions (Megahertz = MHz). All these waves are proceeding at the same speed – that of light – 3×10^8 metres per second. A simple sum therefore will give you the length of each wave.

Divide either frequency or wavelength into 300,000 (for kilohertz) or 300 (for Megahertz) to find the other. Frequency and wavelength are inversely proportional, thus:

$$\frac{300{,}000}{\text{Frequency} = 200 \text{ kHz}} = \text{Wavelength} = 1500 \text{ metres}$$

(BBC Radio 4)

A wide range of frequencies and their corresponding wavelengths are used for different radio and telecommunications applications. For convenience, these frequencies are grouped into *bands* as follows:

Abb.	*Designation*	*Frequency*	*Wavelength*
VLF	Very Low Frequency	3–30 kHz	100–10 km.
LF	Low Frequency	30–300 kHz	10–1 km.
MF	Medium Frequency	300–3000 kHz	1000–100 m.
HF	High Frequency	3 MHz–30 MHz	100m.–10m.
VHF	Very High Frequency	30–300 MHz	10m.–1m.
UHF	Ultra High Frequency	300–3000 MHz	1m.–10cm.
SHF	Super High Frequency	3–30 GHz	10–1cm.

In comparison with each other these waves have somewhat different properties. The longer waves (10 metres and above, that is VLF, LF, MF and HF) can bounce off the charged-particle layers in the upper atmosphere, producing *skywaves* which will travel

Figure 48 VHF and UHF transmitting and receiving aerials are mounted on these masts. The uppermost dipoles are for VHF, and the lower cone-shaped aerials correspond to UHF wavelengths

long distances around the bulge of the earth's curvature. Waves shorter than 10 metres (VHF and above) will only travel more or less along lines-of-sight, and are useless for long-distance applications. Waves which are only a few centimetres long will reflect off small objects, and thus are ideal for radar. They can also be concentrated easily into narrow beams, as with microwaves, which are only a centimetre or two in length.

The design of the aerial depends to a large extent on the wavelength. The optimum length for most applications is a half-wavelength *dipole*. Thus for long waves we need long aerials, long wires for LF and MF, tall poles for HF, aerials about a metre long for VHF, and a few centimetres long for UHF. Microwaves will travel in hollow tubes called waveguides, and can be beamed from the familiar dishes and horns seen on microwave towers up and down the country.

Aircraft Radio Telephony (R/T)

The first and foremost use of radio aboard an aircraft is to communicate with the ground. In the bad old days wireless telegraphy was about the only technique possible, and long wavelengths were used at first. LF, then MF and HF sets were carried by aircraft. The equipment was heavy, bulky, difficult to use and only moderately reliable. Because the signals echoed off the sky, there was a good deal of interference and background noise. Round about the time of World War Two VHF began to be used for aircraft R/T. This cut out interference and static almost at a stroke, and after the War, by international agreement, that part of the VHF band from 118 to 136 MHz was allocated to civil aviation. Military aircraft today prefer to use frequencies in the UHF band, around 300 MHz, partly for technical, partly for security reasons.

HF radio is still used, however, for long-distance flying, over the oceans and well out of range of land. With the help of the skywaves it can produce, HF radio will reach round the bulge of the earth and link planes over mid-Atlantic with the shore-stations at Gander, Prestwick and Shannon.

The frequency band most likely to interest aviation enthusiasts is the civil VHF air-band from 118 to 136 MHz. This is partly because it is possible to listen-in (illegally) using one of the many air-band portable radios now on the market.

Although VHF transmissions produce no skywaves, the line-of-sight range is adequate for most purposes, being better than 200 miles between a ground station and an aircraft at 20,000 feet. The limited range also cuts out much interference and

VHF Aeronautical Radio Frequencies

All frequencies in Megahertz (MHz)

Airport	*Tower*	*Approach, Radar, Director*	*Airways or Zone ATC*	*ILS Localiser Frequency*	*Ident*
Aberdeen	118.1	120.4	131.3, 133.2	109.9	I-AX
Belfast	118.3	120.0, 120.9	128.05, 133.05	109.7 110.9	I-AG I-FT
Birmingham	118.3	120.5	120.5	110.1	I-BM
Bournemouth	125.6	118.65, 119.75	134.45	110.5	I-BH
Bristol/L'gte	120.55	127.75, 124.35	132.8		
East Midlands	124.0	119.65, 120.15		109.9	I-CA
Edinburgh	118.7	121.2, 124.25	121.2	108.9	I-VG/TH
Exeter	119.8	128.15			
Glamorgan/Rhoose	121.2	125.85, 120.05	131.2, 132.6	108.9	I-RS
Glasgow	118.8	119.1	126.25, 124.9	109.3 110.1	I-OO I-UU
Isle of Man	118.9	120.85	128.05	110.1	I-RY
Jersey	119.45	118.55, 120.3	125.2	110.3 110.9	I-DD I-MM
Leeds/Bradford	120.3	123.75, 121.05		110.9	I-LF
Liverpool	118.1	119.85, 118.45	125.1, 133.05	111.7	I-LQ
London/Gatwick	119.45 121.95	119.6, 118.6, 118.95		110.9 110.9	I-GG I-WW
London/Heathrow	See Page 21.				
Luton	120.2	129.55, 128.75		109.1	I-LJ
Manchester	118.7	119.4, 121.35	124.2, 125.1	108.9 109.5	I-DD I-JJ
Newcastle	119.7	126.35, 118.5	131.05	110.5	I-NC
Norwich	118.9	119.35, 124.25		110.9	I-NH
Prestwick	118.15	120.55, 119.6	134.3	110.3 109.5	I-KK I-PP
Stansted	118.15	126.95, 123.8	125.55	110.5	I-SX
Teesside	119.8	118.85, 128.85		111.3	I-TD

Emergency VHF frequency	121.5	London Volmet (South)	128.6
London Volmet (North)	126.6	Shanwick Oceanic	123.95, 127.65

static. Voice communication using VHF is crisp and clear with very little background noise. In fact the R/T equipment installed in today's aircraft is so easy to use that there is no need to carry a specialist radio operator.

The pilot tunes his radio to the frequency of the controller he requires simply by turning a tuning knob through clickstops until the frequency shows in the window. A combination of quartz crystals has now been brought into circuit which will keep the set resonating accurately at this frequency and at no other. This *crystal-controlled tuning* as it is called is clearly preferable to the method of tuning we are used to on our home radios, whereby we twiddle the tuning knob and listen for the best results we can get.

The transceiver is now tuned in. Through the cabin loudspeaker, or through his headphones, the pilot can hear the controller talking to other aircraft. He waits for a suitable gap in the exchanges, then presses the microphone button. He is now transmitting.

'London Control, this is Golf Bravo Alpha Oscar Echo.'

He waits to hear the controller reply, 'Oscar Echo, Go ahead.'

The radio can be switched through the headsets, hand-held microphone, cabin speakers, etc. There are a number of sophisticated arrangements. The microphone button is often mounted on the control column convenient for the pilot's thumb, and when this button is not being pressed to transmit, the entire system doubles as an intercom between the two pilots and anyone else who is plugged in via a headset. The various radios etc., can be switched through this audio system by means of the *Audio Station Box*.

Frequencies and Channels

There is a tremendous demand for channels or spot frequencies within the band from 118 to 136 MHz. Even though 720 channels are available, each separated from the next by no more than 25 kHz, every airfield tower, approach and radar controller, ATC zone or airways sector controller, airline or operating company, etc., has to be on a different frequency unless they are out of range of each other. For example Heathrow and Manchester control towers are on the same frequency – 118.7 MHz – because their transmitters need only have a short range.

It is not possible to list a full range of frequencies in a publication of this type. The information could hope neither to be complete nor up to date. To help enthusiasts get started we list some of the frequencies used by some of the country's more important civil airports. We recommend that you obtain a flight guide such as the RAF En Route Supplement; Northern Europe and North Atlantic or some similar publication (see page 51).

The international VHF emergency frequency is 121.5 MHz. A continuous listening watch is kept on this channel. On UHF radio, as used by military aircraft, the emergency frequency is 243 MHz.

While flying nowadays, especially under IFR (Instrument Flight Rules), a pilot may have to change R/T frequency a great many times. One gadget which has been devised to help the Pilot Who Has Everything Else is a *Digital Frequency Management System*. This is a push-button memory arrangement like a small calculator. The pilot uses it to store beforehand all the frequencies he is likely to use, and can then recall and tune them one-by-one as he flies.

Radio Jargon

Jargon was very necessary in the bad old crackly days before the coming of VHF. There was a lot of background noise, and procedures were devised to ensure that mistakes and misunderstandings would be kept to a minimum.

The most useful procedure is probably the *International Phonetic Alphabet*. It will pay you to learn this because you can use it yourself when spelling out names and addresses, postcards, car registrations etc., on the telephone. Pilots and controllers use it to pass registrations, callsigns, instructions about reporting points etc.

The *callsign* is usually the aircraft's registration if it is a not on a scheduled airline service. If it is on a scheduled service, the callsign will be the flight number, preceded by some codeword, such as 'Speedbird' in the case of British Airways, 'Clipper' for Pan Am etc. For military aircraft callsigns see Chapter Eight.

Messages are kept short and to the point. There is no time for gossip or cheery pleasantries, which is not to say that all pilots and controllers are dull and humorless. They have too much to concentrate on, too much to do most of the time. Listen to the Heathrow Tower controller, who manages to say all of this without pausing for breath as he handles two runways simultaneously. 'Air France 261 airborne at two-zero. Air Ceylon 034 cleared to land Runway Two-Eight Right. Wind three-one-zero, twelve knots. Speedbird 624 line up and hold.' This last phrase instructs the pilot of BA 624 to turn out onto the runway and line up ready to roll. He should not take off yet, but hold – that is, park with his brakes on until cleared to go. The stock phrases in everyday use are:

'Roger' = I have received all of your last message
'Wilco' = I will comply
'Say again' = Please repeat
'I say again' = I repeat, correctly this time
'Affirmative' = Yes
'Negative' = No

A number of phrases from the Q-code have been retained, each with its own special technical meaning.
QDM = Track to steer to ground station, magnetic
QTE = True bearing *from* ground station
QNH = Altimeter pressure setting in millibars to read height as measured from sea level
QFE = Altimeter pressure setting in millibars to read height as measured from the airfield or runway threshold
QSY = Change frequency

Other technical phrases in common use are:
'Special VFR = Special arrangements permitting a VFR flight in controlled airspace (see Chapter Six)
'Maintain VMC' = Keep clear of cloud
'Squawk' = Instructions to a pilot to identify himself on radar by dialling a special code on his transponder
'Squawk Ident' = Use the identification device on the transponder
'Squawk standby' = Switch transponder to standby position
'Squawk Altimeter' or 'Squawk Mode Charlie' = Switch transponder to show altitude readout on radar screen
'Hold' (in the air) = Fly round beacon using the race-track shaped holding pattern (see Chapter Seven)
'Hold' (on the ground) = Stop taxiing and park, brakes on
'Line up and hold' = Taxi onto the runway, lined up to take off, then park until instructed to roll
'Established' = Locked onto the radio beams of the Instrument Landing System
'Direct' = A routeing which omits intervening beacons or other reporting points
'Straight-in' = A landing approach by the shortest feasible route
'Outer Marker' = Pilot reports that he is over the outer marker beacon of the ILS, usually about 4 miles from the threshold
'Heavy' = Aircraft laden close to maximum flying weight
'Vortex' = Large wide-bodied aircraft with turbulent wake

International Phonetic Alphabet and Morse Code

A	Alpha	· -
B	Bravo	- · · ·
C	Charlie	- · - ·
D	Delta	- · ·
E	Echo	·
F	Foxtrot	· · - ·
G	Golf	- - ·
H	Hotel	· · · ·
I	India	· ·
J	Juliet	· - - -
K	Kilo	- · -
L	Lima	· - · ·
M	Mike	- -
N	November	- ·
O	Oscar	- - -
P	Papa	· - - ·
Q	Quebec	- - · -
R	Romeo	· - ·
S	Sierra	· · ·
T	Tango	-
U	Uniform	· · -
V	Victor	· · · -
W	Whiskey	· - -
X	X-Ray	- · · -
Y	Yankee	- · - -
Z	Zulu	- - · ·
1	Wun	· - - - -
2	Too	· · - - -
3	Tree	· · · - -
4	Fower	· · · · -
5	Fife	· · · · ·
6	Six	- · · · ·
7	Seven	- - · · ·
8	Ait	- - - · ·
9	Nin-er	- - - - ·
0	Ze-ro	- - - - -

'Zulu' = Greenwich Mean Time (GMT)
(English is the official language of Air Traffic Control and GMT is in use world-wide.)

Air-band Receivers

You will not hear much of the exchanges described above unless you spend a lot of time on aircraft flight decks or visiting air traffic control units. A third method is to listen in on an air-band receiver. Portable radios which can receive on the 108–136 MHz band

Figure 49 Aviation enthusiasts with air-band radios

are widely available, but under UK law to listen intentionally to non-broadcast transmissions, such as police, fire brigade, air traffic, etc., is illegal. This would be an offence under the 1949 Wireless Telegraphy Act for which the penalty on conviction could be a fine of £200. Despite this these radios are widely used by aviation enthusiasts. Performance is surprisingly good. Aircraft can be picked up over a wide radius, and depending on reception conditions such as geography and weather effects, ground transmitters can be heard also. Most air-band radios will tune down to 108 MHz which will allow you to tune the callsigns of navigation and landing aids. Besides the air-band, they can be switched to one or more broadcast bands as well. Typical of the breed is the little Sharp FX-213 AU which, although it is small enough to slip into a pocket, briefcase or camera bag, performs just as well as its larger counterparts.

The fact is that air-band listening, or air-radio watching, though clandestine, has become a hobby in its own right practised by large numbers of aviation enthusiasts. Unusual or interesting aircraft can be 'spotted' as they call in on radio, and the 'sighting' recorded for the club magazine. If one of these planes lands, the spotters can be at the aircraft fence and brandishing their cameras within minutes of touchdown.

The Legal Position Put very briefly, the legal position is that listening in to aircraft is an offence punishable at law in the UK. If you are found listening in you risk prosecution and a heavy fine. It would be no defence to argue that others do it, e.g. hundreds of kids every weekend at their local airport, or that the airport authorities seem to condone the practice, or that you did not know it was an offence. The law is the law and will remain so until somebody changes it. At the moment there is no pressure group representing the interests of all aviation enthusiasts, which might perhaps lobby for a change in the law, or special dispensations for bona-fide enthusiasts. The 1949 Act may be something of a dinosaur, framed in the days of bulky valves and outdoor aerials, but it still serves a purpose in providing an element of security and confidentiality to the licensed users of non-broadcast radio, as well as a small measure of protection against interference.

The best case for changing the law would be to legitimise or regularise a practice which is already widespread. There are strong educational arguments for permitting air-band listening, perhaps with certain built-in controls, such as licences which might have to be shown to police and airport officials. However, until there is a change the best advice I can give aviation enthusiasts would be to observe the law rather than to ignore it.

If you can't be good be careful. If you must listen in, do it discreetly. Don't let your radio blare out in an airport lounge or snack bar. This is inviting attention. Use the earphone instead. Even out of doors you will find the earphone useful, especially if a Rolls Royce Spey is starting up nearby.

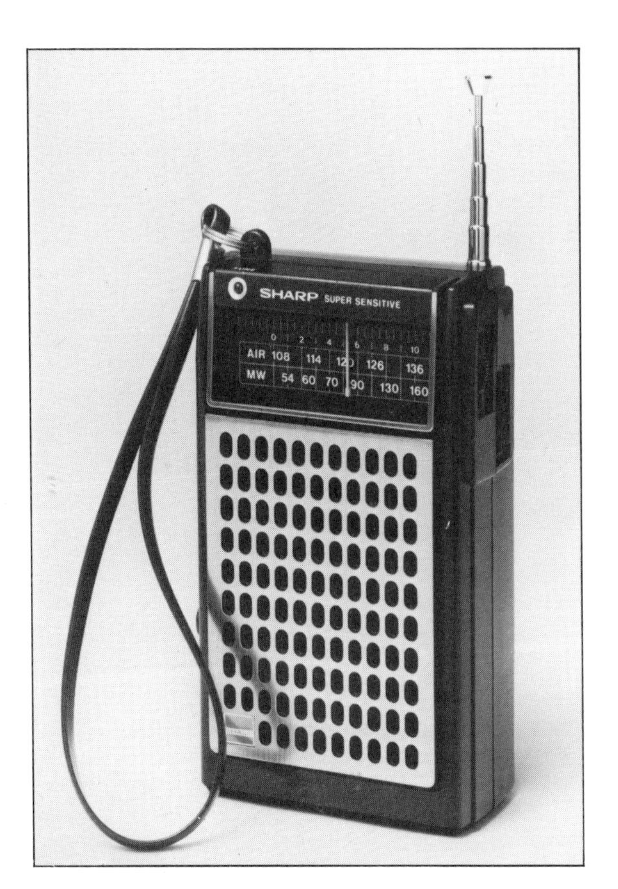

Figure 50 The Sharp FX-213 AU pocket-size air/MW radio is popular with aviation enthusiasts. *Sharp Electronics, U.K., Ltd.*

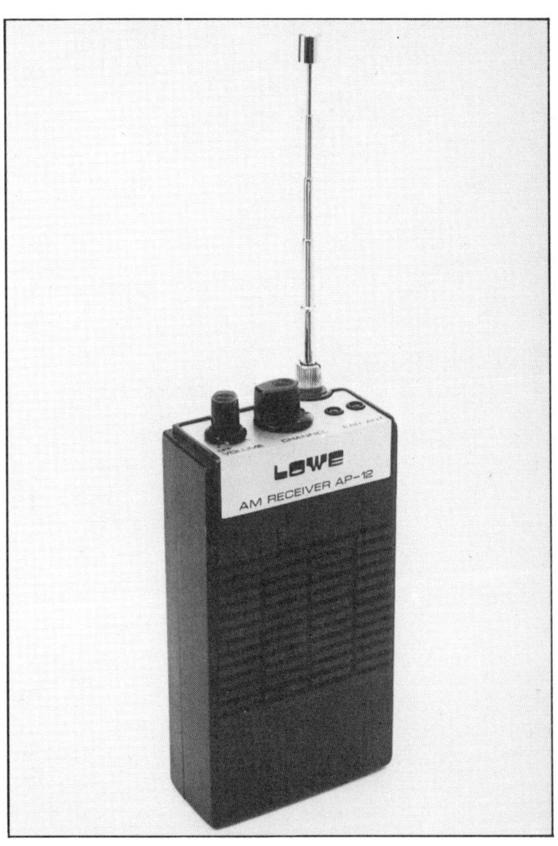

Figure 51 Professional air-band monitors such as these incorporate crystal-controlled tuning. For use by airport personnel, flying clubs, etc. *Lowe Electronics Ltd., Matlock*

Figure 52 'Uniform Juliet. Call London Control now on Wun Tree Tree Decimal Seven'. The Piper Aztec is popular with air taxi operators. *Keenair*

Things to Do

If you can manage to obtain an air-band radio and if it ever becomes legal to use it, a variety of interesting projects can be undertaken. We repeat our warning concerning the illegality of listening in on these frequencies as the law stands at present. If you cannot wait for a change in the law, then at least be discreet; no listening in airport lounges, use ear-phones, try not to flash your radio around.

1 Visit a busy airport and try tuning to each of the following:
   Callsign of Instrument Landing System (ILS), usually around 108–112 MHz
   Callsign of Terminal VOR beacon (110–117 MHz)
   Ground Control (usually around 121–122 MHz)
   Tower Control (usually around 118–119 MHz)
   Approach Control (usually around 119–120 MHz)
   Automatic Departure Information (ATIS) (around 122 MHz)
   Air Traffic Control (124 MHz upwards).
2 Try tuning to VOR beacons between 108 and 112 MHz. Decipher the Morse callsigns and refer to an Aerad map to discover which ones they are. From local high ground or a tall building, how many VORs can you hear?
3 Follow an aircraft on the radio from the time the controller allows the pilot to start up and push back, as it changes frequencies until it takes off and climbs away.
4 Follow an aircraft on the radio from the time it reports in to ATC, down through Approach, Tower and Ground Control until it is parked on the ramp.
5 Use a stopwatch to time an aircraft when it reports itself over the Outer Marker on final approach until it touches down. If you can find the distance to the marker you can calculate (e.g. on a pocket calculator) the average speed of the aircraft during its approach.
6 Collect QNH values off your radio and compare them with the weather you have been having. Compare them also with the figures in millibars shown on television and newspaper weather maps, which because they are corrected for mean sea level, are equivalent to QNH.
7 If you can connect your radio to a tape recorder by means of a suitable lead with DIN plug etc., you can make an air traffic control tape recording. An open-reel tape recorder will be best; one that can play through its speakers when it is not recording, with a quick-release pause switch to eliminate the silences.
8 Use a map of the area round the airfield to plot the tracks of aircraft both outbound and inbound.

5-Beacons and Radio Navigation

Aircraft today navigate mostly using radio beacons of one kind or another. The simplest radio beacons are just ordinary transmitters on which the pilot can get a bearing by using a direction-finder. In fact transmitters operated by the BBC or the GPO can be and often are used for this purpose. An example is the powerful BBC transmitter at Wychbold, near Droitwich, which many pilots find useful as it is powerful enough to be received easily anywhere in UK airspace and in fact deep inside Europe.

However, the purpose-built radio beacons installed by the National Air Traffic Services (a joint civil and military body) throughout the UK are nothing like as powerful as Droitwich. Their power and their range is limited. These beacons are of several kinds. The simplest are the *NDBs – Non-Directional Beacons*. These are small automatic transmitters which radiate a continuous-wave signal uniformly in all directions, day and night, winter and summer. They use LF/MF frequencies. The transmitters are housed in small, usually brick-built huts, and the aerials are either of the tall pole, or of the 'clothes'-line' variety.

A much more sophisticated beacon is the type known as the *VOR* (*VHF Omni Range*). Some of these are like dustbins in appearance; others are fairly large elaborate structures looking like platforms made of girders and wire mesh. The military equivalent of VOR is the network of *TACAN* beacons, which again look like dustbins. Add to this the facility known as *DME* (*Distance Measuring Equipment*), fan markers and the transmitters used for systems such as Decca, and the picture is one of a varied assortment of radio facilities (Navaids) dotted throughout the country. These are positioned in the vicinity of airfields, along approach flightpaths, along the airways, wherever they cross or change direction. They can be used for position-fixing by a pilot flying on DR, but the normal technique is to beacon hop, that is, to fly from one to the next by homing straight onto each.

Locating Beacons

The Civil Aviation Authority's COM 2-0-3 Radio Facilities Chart shows the location of almost all the radio beacons in the UK. The exact geographical positions are detailed in flight guides such as the UK Air Pilot and the RAF En Route Supplement (Northern Europe and North Atlantic) and are also shown on Aerad charts. These positions are given as co-ordinates of latitude and longitude correct to one second of arc. The positions of beacons and other facilities are also given on some radio navigation charts such as the AERAD series.

If you want to go looking for radio beacons in your locality, you may find a pilot or some officer at the local airfield who can help you. Otherwise you will need to do a spot of detective work. Try walking the extended centreline of your local runway, armed with an Ordnance Map, for up to five miles either way. To find the airways beacons however, dotted as they are throughout the country, you will need some guide to the positions of these beacons. If you purchase an AERAD RUR1/2 chart from International Aeradio Limited, Hayes Road, Southall, Middlesex, or the RAF En Route Supplement from No. 1 AIDU, RAF Northolt, West End Road, Ruislip HA4 6NG, you will have the positional details you need. You now plot these positions on a suitable map, such as a One Inch or 1:50,000 series Ordnance map or a four-miles-to-the inch Topographical Air Chart.

Having pinpointed the beacons, you will find that they are not all convenient to the public road. Many are sited on private land, sometimes miles from anywhere, and their only link with civilisation is the track used by NATS vehicles on their regular servicing calls. Take care not to trespass. If you are challenged remain patient and polite.

The only notice posted on most beacons is one which reads: DANGER HIGH VOLTAGE, or in some cases WARNING INTERFERENCE WITH THIS EQUIPMENT MAY ENDANGER AIRCRAFT IN FLIGHT. There will be a transmitter hut with the exhaust stack of an auxiliary generator sticking out of the wall. Not all beacons have generators which start up in the event of the public power supply failing; a number use float batteries instead.

Beacons are usually referred to by name, a name derived from some nearby town, village or local feature. Based on the name is a two- or three-letter callsign which is used to identify the beacon. Every few seconds the beacon transmits this callsign or *ident* in Morse Code. Thus the Lichfield beacon has

Figure 53 A non-directional Radio Beacon. *NDB*
Figure 54 Sketch map showing locations of main NDB beacons

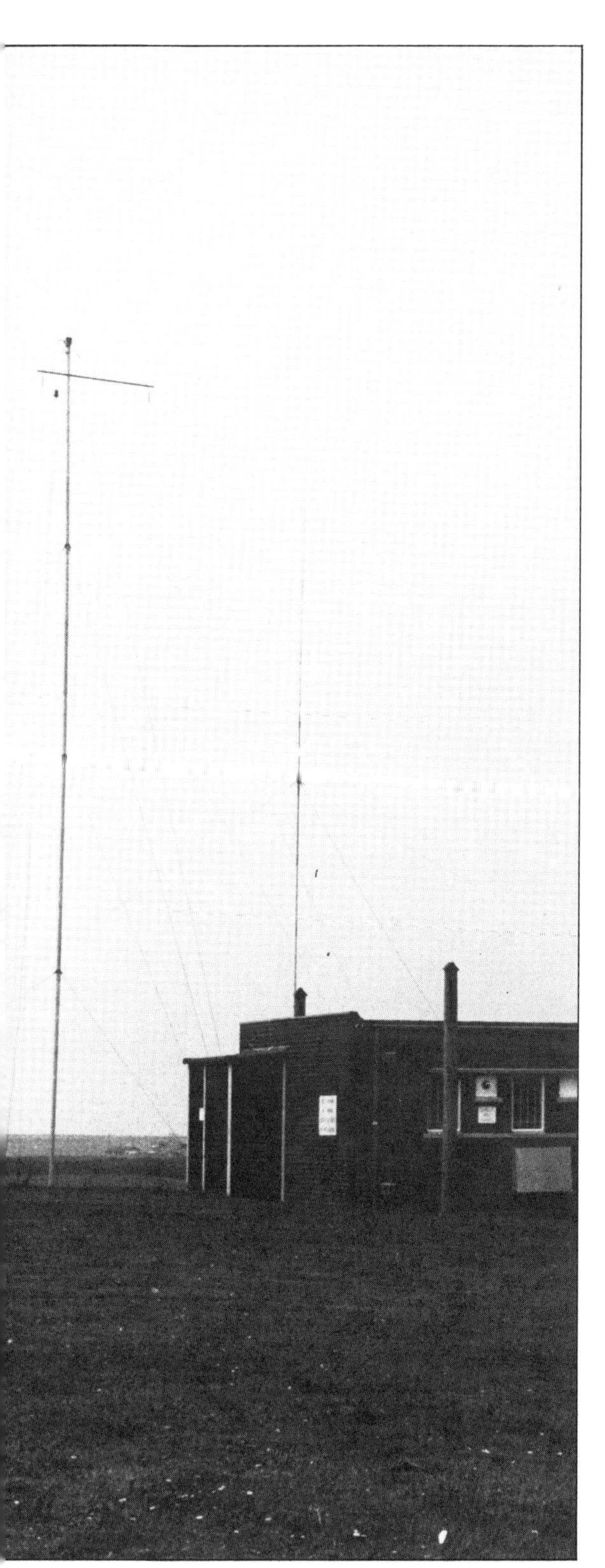

the ident LIC, which it transmits on its own spot frequency of 545 kHz. This is equivalent to a wavelength of 550 metres to which you can probably tune your ordinary domestic or car radio. On a good set you should be able to pick up this callsign anywhere in the Midlands. (Blip-bleep-blip-blip blip-blip bleep-blip-bleep-blip.)

You may find that you can tune to the callsigns of other local beacons, particularly if your receiver will tune down to 1000 metres or below on the LW band. Most beacons of the NDB type in Britain have been allocated spot frequencies between 200 and 600 kHz (1500–500 metres), so you may be able to find them if you scan carefully and slowly across both LW and MW tuning dials listening out for any Morse blips or bleeps. On an air-band radio you can tune to the

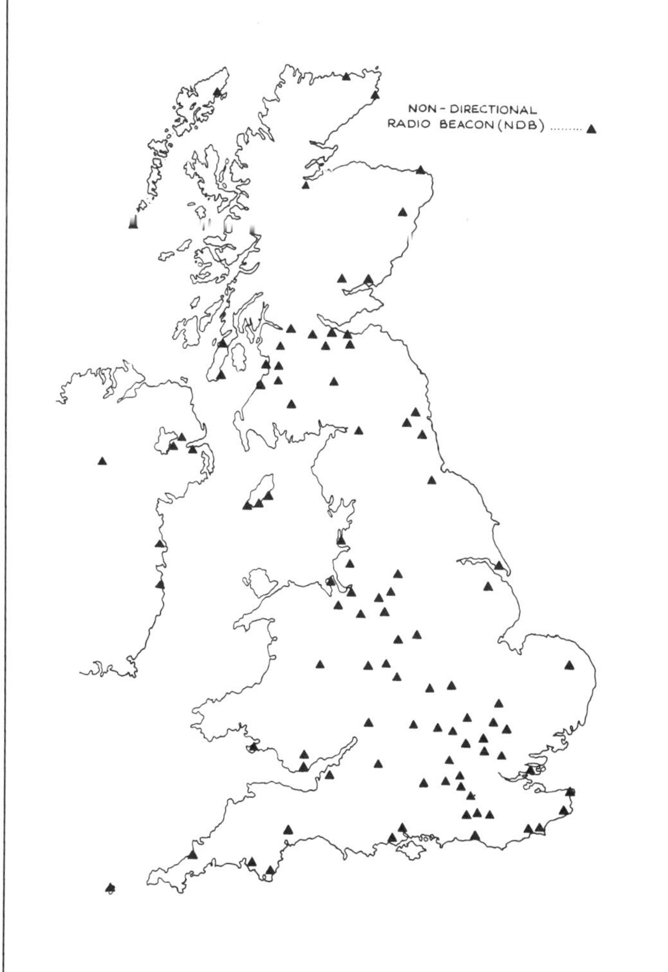

Directory of UK Aeronautical Radio Beacons

VOR = VHF Omni-directional Radio Range (Frequencies in MHz)
DVOR = Doppler VOR (Frequencies in MHz)
DME = Distance Measuring Equipment (UHF Channel shown)
TACAN = Tactical Air Navigation System (UHF Channel and paired VHF frequency shown)
NDB = Non-directional radio beacon (Frequencies in kHz). NDBs are very numerous and only a selection of these is listed.
VB = Voice Broadcast (Automatic Terminal Information Service)

This list may be used in conjunction with the sketch maps on pages 65, 72 and 81

Callsign or Ident	*Name*	*Types of Facility*	*Frequencies kHZ*	*Frequencies MHz*	*UHF Channel*	*Geographical position (of main installation only)*	
ADN	Aberdeen	NDB/DVOR/DME/VB	377	114.3	90	571839N	021557W
ADR	Belfast/A'grove	NDB	346			543735N	061705W
AGV	Aldergrove	TACAN		117.1	118	543923N	061236W
ALC	Alconbury	TACAN		109.0	27	522236N	001250W
ALD	Alderney	NDB	383			494235N	021154W
BCN	Brecon	VOR/DME		116.3	110	514325N	031541W
BDN	Boscombe Down	TACAN		108.2	19	510849N	014512W
BDY	Brawdy	TACAN		116.6	113	515325N	050757W
BEL	Belfast	VOR		116.2		542937N	055156W
BEN	Benbecula	DVOR/TACAN		114.4	91	572834N	072148W
BHD	Berry Head	NDB/DVOR	318	112.7		502353N	032923W
BIG	Biggin	DVOR/DME/VB		115.1	98	511949N	000211E
BKY	Barkway	VOR		113.4		515944N	000359E
BNN	Bovingdon	NDB/VOR/DME/VB	214	112.3	70	514330N	003251W
BPK	Brookmans Park	NDB	328			514451N	000604W
BTN	Barton	VOR/DME/VB		112.4	71	532730N	022721W
BTW	Bentwaters	TACAN		108.6	23	520734N	012635E
BUR	Burnham	VOR		117.1		513102N	004009W
BZN	Brize Norton	TACAN		111.9	56	514450N	013606W
CFD	Cranfield	NDB/VOR	797	116.5		520426N	003635W
CGY	Coningsby	TACAN		111.1	48	530526N	001003W
CHT	Chiltern	NDB	279			513705N	003035W
CLN	Clacton	DVOR/DME		115.7	104	515054N	010859E
CON	Congleton	NDB	360.5			531148N	021135W
CSL	Coltishall	TACAN		116.5	112	524453N	012045E
DCS	Dean Cross	VOR/DME		115.2	99	544319N	032021W
DET	Detling	VOR/DME		117.3	120	511812N	003556E
DTY	Daventry	NDB/DVOR/DME	249	116.4	111	521046N	010644W
DVR	Dover	VOR/DME		117.7	124	510940N	012126E
EDN	Edinburgh	NDB	341			555843N	031702W
EPM	Epsom	NDB	316			511908N	002215W
FAW	Fawley	NDB	394			505156N	012324W
FFA	Fairford	TACAN		110.5	42	514052N	014747W

Callsign or Ident	Name	Types of Facility	Frequencies kHZ	Frequencies MHz	UHF Channel	Geographical position (of main installation only)	
GAM	Gamston	DVOR		112.8		531652N	005644W
GLG	Glasgow Airport	NDB	350			555528N	042006W
GOW	Glasgow	VOR/DME/VB		113.4	81	554204N	042056W
GUR	Guernsey	NDB/DVOR/DME	361	109.4	31	492617N	023609W
HON	Honiley	VOR/DME		112.9	76	522128N	013928W
IBY	Ibsley	DVOR/DME		114.4	91	505337N	014454W
INS	Inverness	DVOR		109.2		573234N	040225W
IOM	Isle of Man	NDB/DVOR/DME	391	112.2	59	540401N	044545W
ISL	Islay	DVOR		109.4		554107N	061523W
JEY	Jersey	NDB	367			491311N	020209W
KIN	Kintyre	NDB	374			552216N	053152W
KNI	Knighton	NDB	404.5			523225N	031335W
KSS	Kinloss	TACAN		109.8	35	573934N	033201W
KWL	Kirkwall	DVOR		108.6		585736N	025333W
LAM	Lambourne	NDB/DVOR/DME	633.5	115.6	103	513844N	000912E
LIC	Lichfield	NDB	545			524447N	014303W
LKH	Lakenheath	TACAN		110.2	39	522422N	003302E
LND	Land's End	DVOR/DME		114.2	89	500012N	053810W
LON	London	VOR/DME		113.6	83	512912N	002754W
LUK	Leuchars	TACAN		110.5	42	562223N	025144W
LYD	Lydd	NDB/VOR	397	116.7		510001N	005310E
MAC	Macrihanish	VOR		110.2		552624N	054119W
MAY	Mayfield	NDB/DVOR/VB	343	117.9		510100N	000704E
MAZ	Macrihanish	TACAN		116.0	107	552628N	054200W
MID	Midhurst	VOR/DME		114.0	87	510312N	003724W
MLD	Mildenhall	TACAN		115.9	106	522146N	002723E
NEW	Newcastle	NDB/VOR	352	113.5		550149N	014322W
NGY	New Galloway	NDB	399			551038N	041002W
NQY	St. Mawgan	TACAN		112.6	73	502558N	050030W
OCK	Ockham	VOR/DME		115.3	100	511816N	002643W
OLD	Oldham	NDB	344			533318N	020258W
ORM	Ormskirk	NDB	315.5			533537N	025139W
OTR	Ottringham	NDB/DVOR/TACAN	335	113.9	86	534153N	000608W
OUN	Ouston	TACAN		113.5	82	550132N	015258W
POL	Pole Hill	DVOR/DME		112.1	58	534437N	020606W
PTH	Perth/Scone	NDB	388			562603N	032218W
PWK	Prestwick	NDB/DVOR/DME	355	117.5	122	551848N	044700W
SAB	St. Abbs	NDB/VOR	284	112.5		555427N	021217W

Callsign or Ident	Name	Types of Facility	Frequencies kHZ	MHz	UHF Channel	Geographical position (of main installation only)	
SFD	Seaford	VOR/DME		117.0	117	504535N	000725E
SHD	Scotstownhead	NDB	383			573333N	014856W
SKP	Skipness	NDB/DVOR/DME	385	113.0	77	554435N	053147W
SKT	Sculthorpe	TACAN		114.7	95	525036N	004530E
SM	St. Mawgan	NDB	356.5			502647N	050000W
STN	Stornoway	NDB/DVOR	669.5	115.1		581233N	061100W
STZ	Stornoway	TACAN		115.1	98	581248N	062035W
STU	Strumble Head	NDB/VOR/DME	400	113.1	78	515940N	050219W
SUM	Sumburgh	NDB/DVOR/DME	351	117.3	120	595245N	011704W
SXV	Saxa Vord	TACAN		114.9	96	604939N	005016W
TIR	Tiree	DVOR/DME		117.7	124	562936N	065228W
TLA	Talla	NDB/DVOR	363	113.8		552957N	032105W
UPH	Upper Heyford	TACAN		113.7	84	515616N	011530W
VLN	Yeovilton	TACAN		111.0	47	510019N	023819W
VNR	Ventnor	TACAN		108.9	26	503638N	011417W
VYL	Valley	TACAN		108.4	21		
WAL	Wallasey	NDB/VOR	331.5	114.1		532330N	030800W
WAZ	Wallasey	TACAN		114.1	88	532500N	030640W
WCO	Westcott	NDB	734			515109N	005738W
WDB	Woodbridge	TACAN		117.1	118	520459N	011728E
WET	Wethersfield	TACAN		108.8	25	515900N	003000E
WHI	Whitegate	NDB	368.5			531105N	023717W
WIK	Wick	NDB/VOR	344	113.6		582733N	030552W
WIZ	Wick	TACAN		113.6	83	582741N	030456W
WOD	Woodley	NDB	357			512709N	005240W

callsigns of VOR beacons and ILS installations in the sector from 108 to 118 MHz. Listening to and deciphering callsigns is good fun, and is also a good way of learning Morse Code. This is much less difficult than many people think, and once you have become used to a few letters or groups, it is amazing how quickly the pattern of dots and dashes begins to seem as clear and sharp as the letters printed on this page.

Beacons and other radio facilities have a two-fold fascination for aviation enthusiasts. Firstly there is the interest which the navaids themselves arouse as examples of radio or avionics technology. Secondly there is the clear relationship which navaids have to the flightpaths followed by aircraft. These small huts mark out on the land the geography of the air and have an importance for aerial navigation out of all proportion to their size. They are the all-weather signposts and buoyage systems of the air, marking most of the reporting points and other places which aircraft customarily overfly and to which pilots refer when they are thousands of feet above.

ADF

This method of radio navigation relies on fixed transmitters on the ground and a direction-finder aboard the aircraft, and this system has been in use since prewar days. The transmitters on the ground can be either broadcasting stations such as Droitwich, purpose-built aeronautical beacons of the NDB type, or marine beacons, etc. Aircraft engaged in search and rescue can use their direction-finders to locate and home in on any transmitter carried in the lifeboats or on the persons of the survivors.

For navigation purposes, a bearing on any two transmitters can be plotted on a chart to provide a fix or position. More typically a pilot would simply fly from beacon to beacon, as they are strung out along the airways and other regularly-used routes.

The *Automatic Direction Finder (ADF)* installation on a modern aircraft consists of three main parts. First the aerials. These usually take the form of horizontal 'towel-rail' aerials outside the fuselage, or flush-mounted fittings along some non-metallic part of the aircraft's skin. Unlike the old type of DF loop aerials which had to be turned by hand to find the bearing of the transmitter, modern ADF installations resolve the bearing electronically. The second part of the installation is the tuner/receiver. This can be tuned usually with the help of crystals over a range of frequencies from 190 to 1800 kHz. Finally there is the ADF meter, or *Radio Compass* as it is often called. A pointer sweeps a compass card to indicate the bearing of the transmitter relative to the aircraft's heading.

To use the system the pilot consults his radio navigation chart for details of beacons along or adjacent to his route. The chart will give the frequency and callsign and show the approximate location. He tunes the frequency on his ADF receiver and switches this into his headset so that he can hear and check the callsign. By now the pointer of the ADF meter should have swung to point towards the beacon. If he steers to line up with the pointer, he can home in on the beacon and overfly it, and as he does so the pointer will swing round to point astern. The pilot now knows that he has just passed overhead the beacon.

NDBs (Non Directional Beacons)

These are the purpose-built automatic transmitters specially intended to be used with ADF. There are a large number of them scattered throughout Britain, both on and off the airways, and the sketch-map

Figure 55 A pair of HS 125 Dominie navigation trainers of the RAF, based at Finningley. Note the 'towel rail' ADF sense aerials above the rear fuselage. *MOD*

Figure 56 How an Automatic Direction Finder works

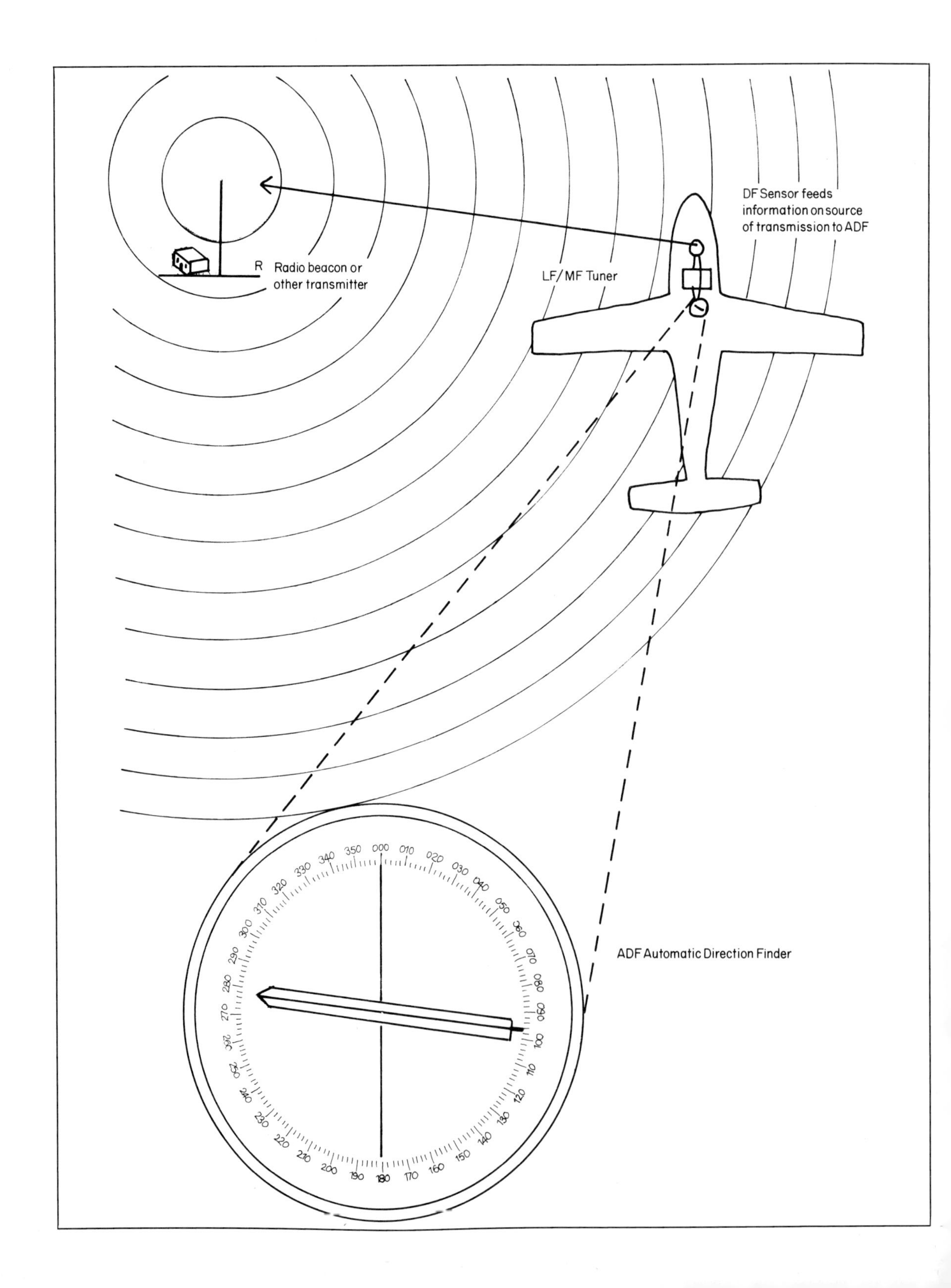

cannot show them all. Because these simple transmitters require no calibration and very little maintenance they are installed as let-down and approach aids, in the vicinity even of quite small airfields. Look for NDBs around or near your own airfield. Examples are GE (Golf Echo) to the east of Gatwick, CA (Charlie Alpha) at Castle Donington and LBA (Ell-Bee-Ay) at Leeds.

NDBs have a simple uninspiring appearance. A hut houses the transmitter and the aerial is usually rigged outside. The aerial is either a tall pole itself, or a wire rigged between two masts. Each NDB has its own frequency, but as the number of available frequencies is limited, interference is kept at a minimum by restricting the power and range of each beacon. The range of most NDBs is 50 n.m. or less.

NDBs installed for marine use need to have a much longer range than this, and the problem of frequencies is solved by grouping about six Marine NDBs together. Using a common frequency, they each transmit for one minute out of every six on a rota system. However, on board a ship there is usually more time available than in an aircraft for dealing with this added complication.

Problems with NDB/ADF Although this system of navigation has been in use for many years and still proves useful today, there are some well-known disadvantages. The first is weather. The LF/MF bands are susceptible to static and other kinds of atmospheric interference, so if there are thunderstorms about, it may be impossible to tune the ADF or the bearings given may not be at all reliable. The ADF may also indicate freak beacons caused by unregulated transmissions. A ghost or freak beacon resulting from power cable transmission is held to be at least partly to blame for the crash of an Invicta Airways Vanguard near Basle in 1974.

The pointer of the ADF will only show the bearing of a beacon in relation to the aircraft's heading. As the aircraft turns the ADF reading will also change, and if the pilot wants the QDM to the beacon, he needs to do some mental arithmetic with one eye on the compass. This problem is overcome with the type of Radio Compass known as an RMI – *Radio Magnetic Indicator* – an instrument in which the compass card is 'live' and rotates against the lubber line to show the aircraft's heading, while the ADF pointer will always show the correct QDM.

To home onto a beacon a pilot may steer to keep the ADF pointer along the lubber line. If there is any crosswind, this must be allowed for. The aircraft is steered a few degrees either to left or right of the beacon – depending on the strength and direction of the crosswind component, otherwise the beacon will be approached not in a straight line, but in a kind of spiralling curve. As the pilot is never sure exactly how much allowance to make for drift, there is always a small element of uncertainty when steering on an ADF.

On the Beam

It is a commonplace that aircraft are guided by radio beams, both en route and during an approach to land. In the layman's mind there is a mental image of a radio beam as a narrow pencil of rays marking some kind of electronic pathway in the sky. This is roughly true. There are radio pathways marked out in the sky, but the trouble with pencil beams, as they are called, is that very narrow beams require either too large an aerial or too short a wavelength – e.g. microwaves, which are impractical for some applications. The other problem with pencil beams arises when an aircraft strays off the beam. The pilot needs some indication that he has strayed left or right so that he can steer to regain the beam, and if there is a single narrow beam this indication is not present – flying such a beam would be a hit-or-miss affair.

This latter problem has been completely overcome with radar-guided weapons of the type known as beam-riders, but these are a relatively recent development and have little to do with day-to-day aeronautics. An earlier solution and one which is still very much with us employs the principle of the *equisignal*, as it is termed. A transmitter radiates signals in two broad lobes in such a way that they overlap slightly. If different signals are broadcast in each lobe, they will then be received with equal strength only along the centreline of the overlap – the *equisignal*. Find some way of marking and detecting this equisignal and you can fly along it. Should you wander either to left or right, you can tell from the modulation which lobe you are in and steer to regain the equisignal. You now have an electronic pathway in the sky – an airway – a track which can be marked as indelibly on the aeronautical landscape as the M1 passing Watford Gap.

The first device to apply this principle in a significant way was the *A–N Radio Range*, developed in the US in the 1920s. This worked as shown in the diagram (fig. 58) by transmitting in four adjacent lobes. These overlapped to form four airways along the equisignals converging on the transmitter. In two of the lobes the Morse letter A (dot-dash) was

Figure 57 Sketch map showing locations of VOR/DME beacons

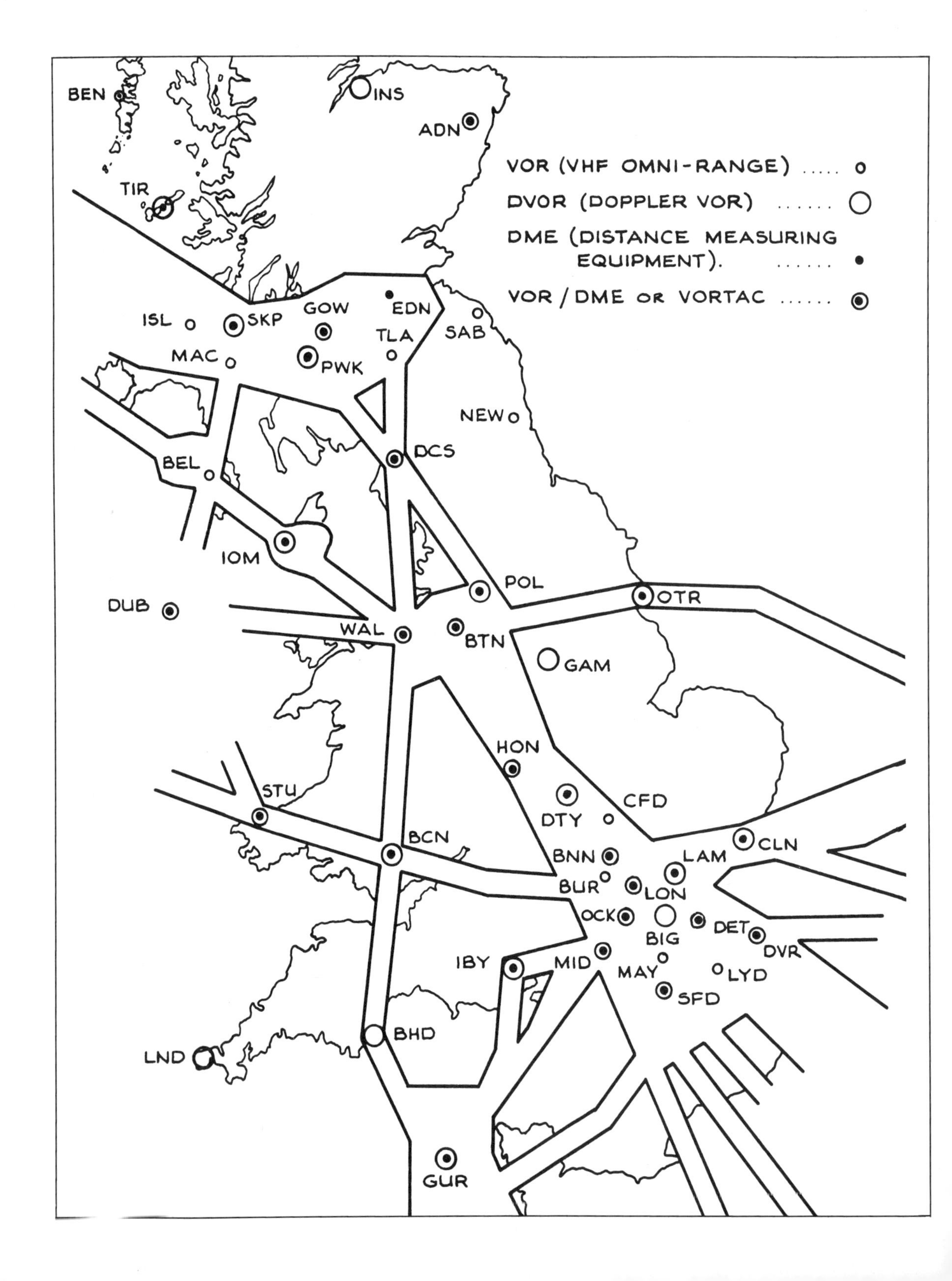

Figure 58 Four airways radiate from a Radio Range transmitter, each defined by the overlap between adjacent A and N lobes

broadcast, in the other two the Morse letter N (dash-dot). These transmissions interlocked in such a way that along the equisignal a continuous tone ought to be heard. A pilot flew the airway by listening through his headset to stay in the equisignal, steering one way if he began hearing A, the other way if he began hearing N. This simple system kept the plane on the straight and narrow even if there was a crosswind – no need to worry too much about compasses or working out the drift angle. Drift takes care of itself, as they say. The first airways in the US, the first roads in the sky, used these radio ranges. However the A–N radio range was rendered obsolete by the arrival after World War Two of the VHF Omni-directional Radio Range, or *VOR*.

However, the principle of the equisignal was incorporated in a number of other landing and navigational devices of the 'beam' variety. In the Standard Beam Approach or Lorenz Blind Landing System dots and dashes were broadcast in adjacent lobes, which blended again into a continuous tone equisignal. This device was adapted by the Germans in 1940 to become the infamous 'Knickebein' blind bombing system. The British, in what is called The Battle of the Beams, devised the countermeasure of laying down false equisignals which got the Heinkels lost over the sea! More peaceful uses for equisignals are in the navigation system known as Consol, and in the familiar present-day Instrument Landing System, or ILS.

The VOR

The VOR is probably the most important of the navigational devices in use today by civil aircraft. It can be considered as an advanced kind of beacon – in fact most VORs are positioned alongside NDBs throughout the airways network. However the name VOR (*VHF Omni Range* or *VHF Omni-directional Radio Range*) suggests comparison with the obsolete A–N radio ranges, because the VOR lays down electronic airways which can be steered regardless of drift. Whereas the old radio ranges could cater for four airways at best, the VOR lays down an infinite number

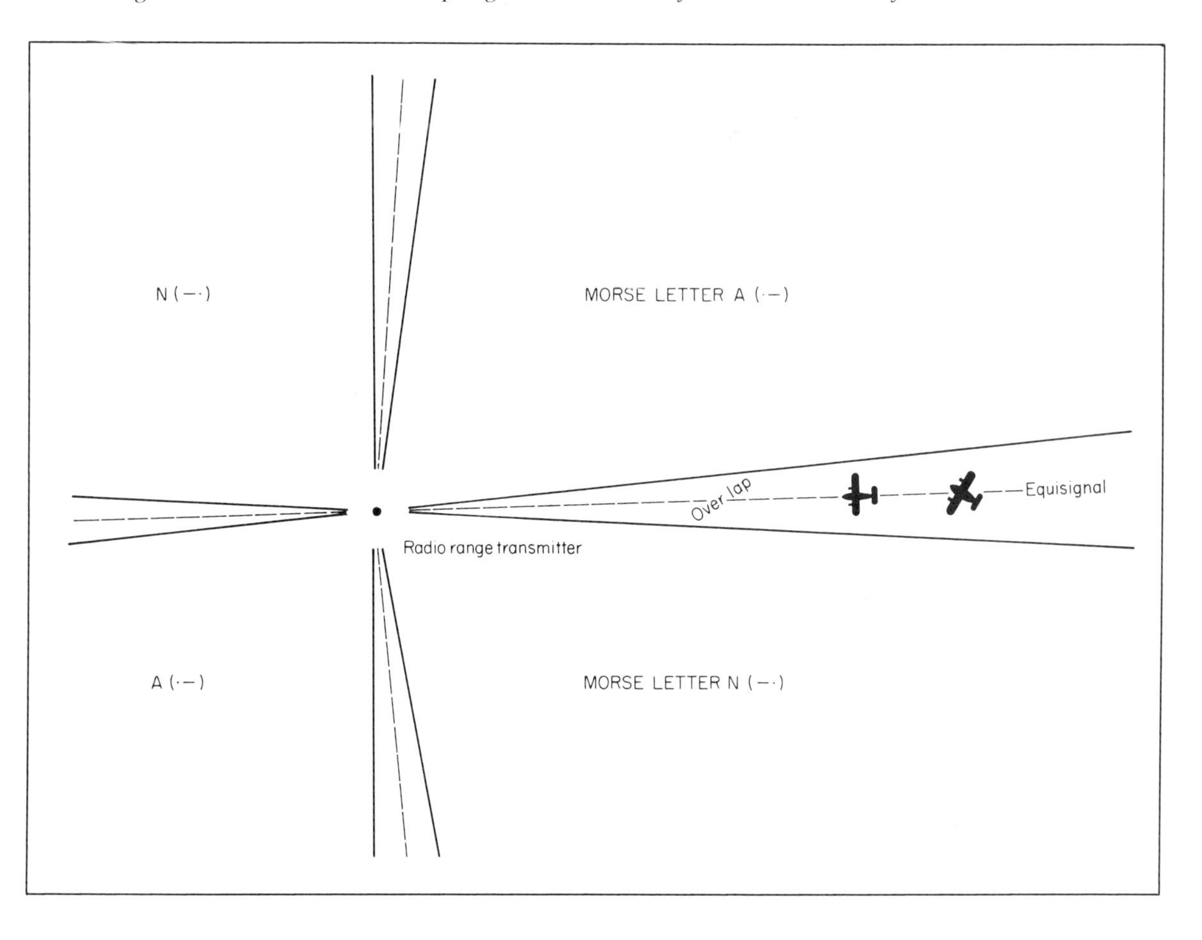

of airways, radials, or beams round all 360° – hence the name omni-directional. Each airway (or *Radial*) is marked by a unique electronic signal which can be decoded by instruments carried in the aircraft, and the same instruments will also warn the pilot should he wander off the beam.

VORs transmit on spot frequencies in the 108 to 118 MHz section of the VHF band. They are mostly free from interference and static and are capable of the full ground-to-air line-of-sight range of up to 200 n.m. No two installations within several hundred miles of each other share the same frequency.

VORs work on the principle of phase comparison. Two signals are broadcast, each modulated in a complex way. An aircraft along any particular radial from the VOR will detect a phase difference between the signals. This phase difference can be stated as an angle, and the beacon is calibrated in such a way that the phase-difference angle (in degrees) is exactly the same as the magnetic bearing of the radial. VORs are tuned on the aircraft's NAV receivers, and the bearing information is displayed on some kind of VOR meter.

Meanwhile, back at the beacon, complex techniques are needed to produce the radially-coded signal. It is necessary to generate a rotating radiation pattern by some means. In the older type of 'dustbin' VORs one of the signals is produced by a directional aerial rotated mechanically at 30 Hz. The newer *Doppler VORs* have no moving parts but use solid-state switching to rotate the signal round a ring of dipoles. Doppler VORs are large, weird-looking installations, which can be calibrated more accurately than the dustbin types.

Most UK VORs are of the dustbin type, but the more accurate Doppler VORs are taking over at many sites, as can be seen from the map. It is worth trying to track down a Doppler VOR; it is a strange and spectacular installation, and it is unlikely that an ordinary layman who happened to come across one in the course of his travels would have even the remotest idea what it was for!

If you have an air-band radio it is probable that it

Figure 59 VOR (VHF Omni Directional Radio Range) beacon. This example is located on the Wirral peninsula near Wallasey on Merseyside. It transmits on 114.1 MHz using the callsign WAL

Figure 60 VOR airway and a VOR indicator meter

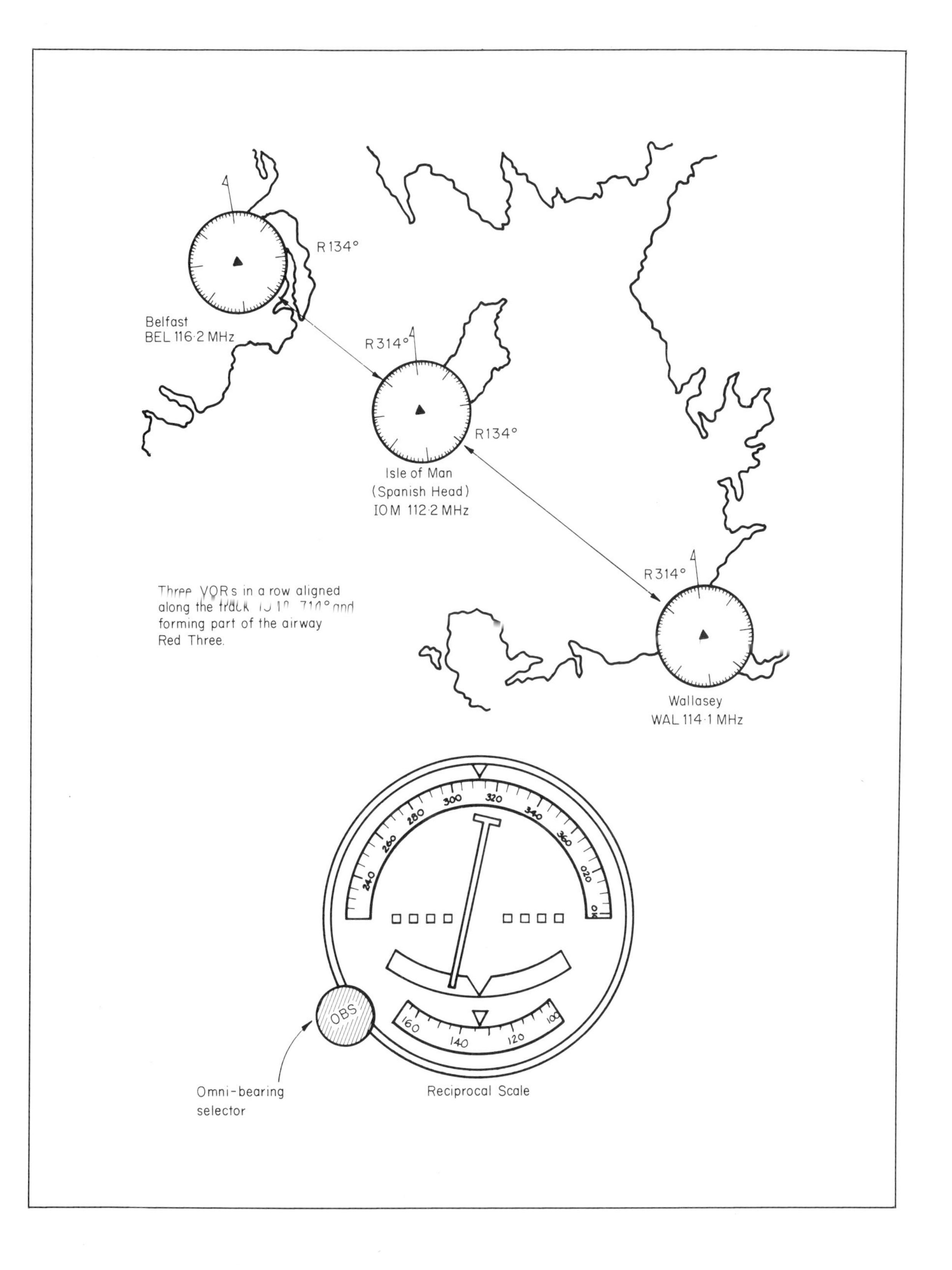

will tune to the callsigns of local VORs for you. You will hear the usual three-letter Morse callsign and a whuffling background noise which is the rotating modulated signal. Some VORs broadcast a pre-recorded voice message containing ATIS arrival information on behalf of the local airport. ATIS broadcasts are available from BNN and BIG for Heathrow, MAY for Gatwick, GOW for Glasgow and ADN for Aberdeen, and although it is on the same frequency this voice message seems not to interfere with the VOR's bearing signal.

When compared to the NDB/ADF system, the VOR operates on a different principle altogether. With VOR, the beacon itself is able to say to the aircraft: 'You are now on such-and-such a radial', whereas with the older system a simple transmitter is sending out an uncoded continuous signal. It is up to the aircraft's ADF to say: 'There is a transmitter out there somewhere and I find it to be about 30° (or whatever) to the left of the plane's present heading.' Thus VOR provides more positive and more accurate bearing information than NDB/ADF.

Meanwhile, on the flight deck, a number of different types of instrument will be used to display VOR information. In a *HSI* (*Horizontal Situation Indicator*) or an *RMI* (*Radio Magnetic Indicator*) the radial will be indicated by means of a pointer, and with sophisticated flight control or director systems, the autopilot can lock onto and fly any selected radial. Dial the next beacon and the plane takes you there. On simpler flight decks the VOR display is usually in the form of an *Omni-Bearing Indicator*. The pilot can rotate a compass ring to select any radial he wishes. As long as he flies to bring the deflection pointer central and keep it there, he is on the radial, and as he overflies the beacon, a little flag clicks to read FROM instead of TO.

DME (Distance Measuring Equipment)

This is a system for measuring the distance from an aircraft to a suitably-equipped beacon, usually a VOR. The VOR will give you a radial, and the DME will tell

Figure 61 This impressive structure is the new Doppler-type VOR which has been installed near the village of Maidford to the south of Daventry. The callsign is DTY on 116.4 MHz. The Daventry facility ranks as one of the most important radio navigation aids in the UK

Figure 62 Radio boxes and navigating instruments show clearly in this composite photograph, taken aboard an Islander on a flight southeast from Belfast along the airway Red Three. Note speed, altitude and heading as shown on the flight instruments. Both DME and VHF NAV 1 are tuned to the VOR at Spanish Head on the Isle of Man as the aircraft flies along the 314° radial toward the beacon. The ADF is tuned to the NDB at Wallasey (331 kHz), the transponder is 'squawking' the identification code 6471, while the communications radios are set to London Volmet weather broadcast and the ATC airways frequencies respectively

you exactly how far out along it you are. This makes precise navigation easy, and for ATC purposes it is possible for a pilot to give an accurate estimate of his arrival time over the beacon. Aircraft need to be fitted with a special black box known as a DME set, and the VOR beacon needs to incorporate a DME *responder*. Many UK VORs are now fitted with DME.

DME has been borrowed (some might say stolen!) from the military TACAN system, and is identical to the distance-measuring component of TACAN. Pilots who carry DME can use it to measure distance either from civil VOR/DME or from military TACAN beacons. At some VOR sites a full TACAN beacon is also installed, giving an arrangement known as VORTAC.

DME is a UHF system using channels consisting of pairs of frequencies in the UHF band from 962 to 1214 MHz. One frequency is used by the aircraft to transmit to the beacon. The other frequency is used by the beacon to reply to the aircraft, and the difference between the pair of frequencies is always 63 MHz either way. This system overcomes for one thing the problem of echoes interfering in either direction.

All the pilot does is to select on his DME set the frequency of the associated VOR or ILS. He is now tuned to the DME responder beacon. The airborne set begins to transmit on one frequency a set of uniquely-coded pairs of pulses. No other aircraft within range should be transmitting the same code. The beacon will store all the codes it can hear and after a brief period retransmit them on the other frequency. Meanwhile the receiving part of the airborne set will be listening out on the new frequency for the first correctly-coded reply. The airborne DME recognises its own pulse-code and measures the time taken for the entire process. This is computed by the set and displayed to show the distance to the beacon correct to 0.2 n.m. Two consecutive measurements of distance can be computed to show the aircraft's speed, and yet another mode of display will show the flying time (in minutes) – very useful for the pilot who must announce his arrival to Air Traffic Control.

Fan Markers

These are installed nowadays mainly as components of Instrument Landing Systems. They are installed along the ILS approach route, directly on the extended runway centreline. The Middle Marker is usually about one mile out, the Outer Marker about four miles out. Markers have been used to indicate a small number of reporting points along the airways, but this practice is being gradually abandoned.

A marker has a directional aerial array consisting of a pair of folded dipoles mounted above the roof of the hut and radiating on 75 MHz. The signal is beamed upwards in a cone or fan, and can only be received by aircraft which are flying through the fan. The fans do not overlap so there is no interference between them, and the same frequency (75 MHz) can be used for all fan markers. Aircraft carry a special marker receiver permanently tuned to this frequency and normally left switched on throughout the flight. When an aircraft enters a fan a series of bleeps can be heard and coloured lights flash on the instrument panel.

	Audio Tone	*Modulation*	*Lights*
Airways Markers	300 Hz	Morse Ident	White
ILS Outer Marker	400 Hz	2 dashes/sec.	Blue
ILS Middle Marker	1300 Hz	Dots and dashes	Amber

VDF

Ground-based direction-finders are still in use nowadays, despite radar. A pilot who has no radio-navigation equipment may call in on R/T and ask for a bearing, usually a QDM or a series of QDMs. The bearing of the aircraft which is transmitting shows up instantly on the small cathode-ray tube of the *VDF* (*VHF Direction Finder*) in the control room on the ground, and this is radioed back to the pilot. Military airfields use UDF systems (*UHF Direction Finder*). A network of special direction-finder stations permanently tuned to the emergency frequency of 243 MHz covers most of the UK. Each is linked by landline to the London Military ATCC at West Drayton, thus enabling any military aircraft in distress to be pinpointed instantly as soon as it calls in on 243 MHz.

Navigation Using VOR/DME

VOR and its counterpart DME, backed-up as required by NDBs and in some cases fan markers, is the main official international system of air navigation for civil purposes. The VOR/DME system was chosen in preference to Decca by the International Civil Aviation Organisation (ICAO), and even if aircraft navigate by other methods they are still controlled by reference to VOR/DME. The operating manuals of most commercial airlines specify that, when carrying passengers, aircraft shall fly in accordance with the *Instrument Flight Rules* (*IFR*). In practice this means complying with instructions from Air Traffic Control, overflying VORs and using DME as required.

Figure 63 The London VOR/DME beacon located just to the north of Heathrow Airport. This gives bearing and range information for aircraft approaching or leaving airfields in the London area. The callsign LON can be heard on 113.6 MHz

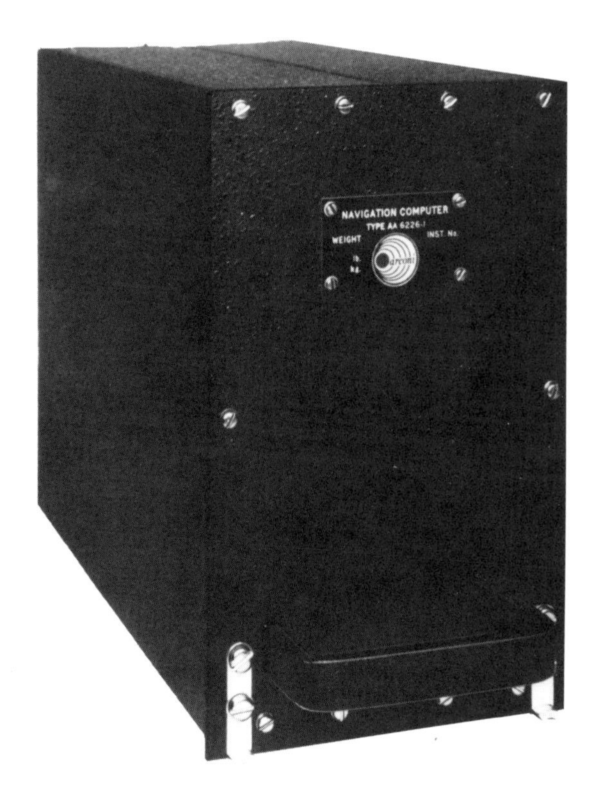

Figure 64 R-NAV navigation computer. *Marconi Avionics Limited*

R-Nav or Area Nav

VOR/DME is an easy way to navigate if you fly along the airways from beacon to beacon. Off the airways it is necessary to work out your position by taking *offsets* from VOR/DME. An offset is a distance along a given radial. It is fairly easy to use offsets to fix your position relative to a chart, but much more tedious to fly a course to any predetermined offset (*waypoint*). Do you intercept the radial first, then fly along it until you are at the correct distance; or do you get the distance on the DME correct, then orbit the beacon towards the radial you want? Either method is primitive when you can have a little computer called R-Nav to do the work for you. Installed in the instrument panel is a small push-button control box looking like a pocket calculator. The pilot pre-selects the *waypoints* he wishes to fly to by defining these in terms of offsets (Radials plus distance) from a suitable VOR/DME or VORTAC. The computer will not only display the aircraft's position continually, it will also feed into the autopilot, flight control and director systems to steer the aircraft to each waypoint in turn. To the pilot scanning his HSI and DME indicators in the normal way it is as though a ghost beacon were positioned at each waypoint. The real beacon, still alive and well and operating happily several miles away, has been mathematically offset by the computer and repositioned at the waypoints!

TACAN (UHF Tactical Air Navigation System)

This is a system of navigation used by military aircraft which relies on ground beacons to provide bearing and distance information on frequencies from 962 to 1214 MHz. The system could be described as the military counterpart of VOR/DME. It can be used in peacetime for general purpose and en-route navigation, homing, let-down and even for instrument landing approaches. It is also accurate enough to be used during hostilities for pinpointing strike targets, but then of course the TACAN beacons themselves would be susceptible to counter-attack.

The bearing information from a TACAN beacon is provided by a rotating aerial housed in a 'dustbin' feature, similar in appearance to a VOR. Phase-comparison techniques are used and the aerial rotates at 15 Hz. The distance-measuring component is the same as that in civil DME. TACAN beacons are usually mounted on top of lattice towers, with a built-in foldaway hoist for servicing the dustbin. TACAN is also compact enough to be transported easily, deployed in forward areas, etc. Mobile TACANs are sometimes used for air displays, as at the 1977 Jubilee Review at RAF Finningley. The locations of all permanent TACANs in Britain are shown on the sketch-map. These are mostly installed at or near military airfields. Only a few are installed on civil airways, and there are no military airways as such. However the TACAN Routes are used by certain

Figure 65 A TACAN beacon

routine military operations, such as sorties to training areas and bombing ranges. Some of the flight-refuelling corridors are aligned along TACAN radials.

Aboard the aircraft there may be a separate TACAN indicator, but it is more usual nowadays to show the radial on the pointer of the HSI compass instrument, with the DME distance shown in counters alongside. *Offset Tacan*, the military equivalent of R-Nav, is fitted to some of the latest aircraft, such as the Jaguar. Lossiemouth Jaguars can get home by offsetting the Kinloss TACAN by 7.7 n.m. along the radial 073°.

The DME component of TACAN (but not the bearing information component) can be used in an air-to-air mode, for example between a pair of aircraft. They set their TACANs to channels which are 63 MHz apart, for example Channels 29 and 92 – a combination which is easy to remember. The DME in each aircraft now shows the distance to the other.

Radio Navigation Charts

These are charts which contain all the information necessary for flying on radio instruments. Beacons of different types are shown, NDB, VOR, DME, TACAN etc., together with the details of frequencies and callsigns, radials and distances along airways etc. Hardly any topographical detail is shown apart from coastlines. The keen air watcher and aviation enthusiast will find these charts of absorbing interest. There should be no need to buy charts for yourself unless you particularly want up-to-date ones. Pilots, flying clubs, etc. throw away their old charts as soon as new ones are published.

There are a number of publishers of these charts, and as they are in competition they tend to vie with each other in devising systems for getting as much information as possible into the smallest area or weight of paper. Some pilots tend to favour the charts published by the American firm Jeppesen. The UK publishers are International Aeradio Limited, a subsidiary of British Airways, and the RAF. Both of these provide very informative and compact charts of UK airspace, and also of airspace throughout the world.

We publish an extract here from the London Area chart of the AERAD series, together with a key to the symbols in the form of an AERAD Legend Card. The AERAD series of charts is available from British Airways Flight Services, Bealine House, PO Box 7, Ruislip, Middlesex, HA4 6QL. The combined EUR 1 and EUR 2 chart shows all the radio aids in the UK.

The radio navigation charts published by the RAF are known as En Route Charts. Best for the UK is ERC 412N-L which is backed by the ERC 412S-L. The High Altitude Chart for the British Isles and North Sea is ERC 411-H, and the chart covering South East England to an enlarged scale is ERC 425T-L. All these RAF charts are available for sale to the public and are available from: No. 1 AIDU, RAF Northolt, West End Road, Ruislip, HA4 6NG.

Other Navigational Techniques

A full description of Decca, INS and other navigational systems used by modern aircraft is outside the scope of this book; however, a brief account may prove useful to the reader who wishes to appreciate these systems in the context of those already described. Basically there are two broad types of navigation system which it is possible to devise. One type uses ground transmitters similar to beacons, whereas other systems are fully airborne and independent of ground transmitters. Such systems can not only be used anywhere as a general-purpose aid to navigation; they also have a great tactical importance, for example as a blind bombing aid, capable of accurately pinpointing targets in forward areas.

Figure 66 TACAN beacons and associated military air routes

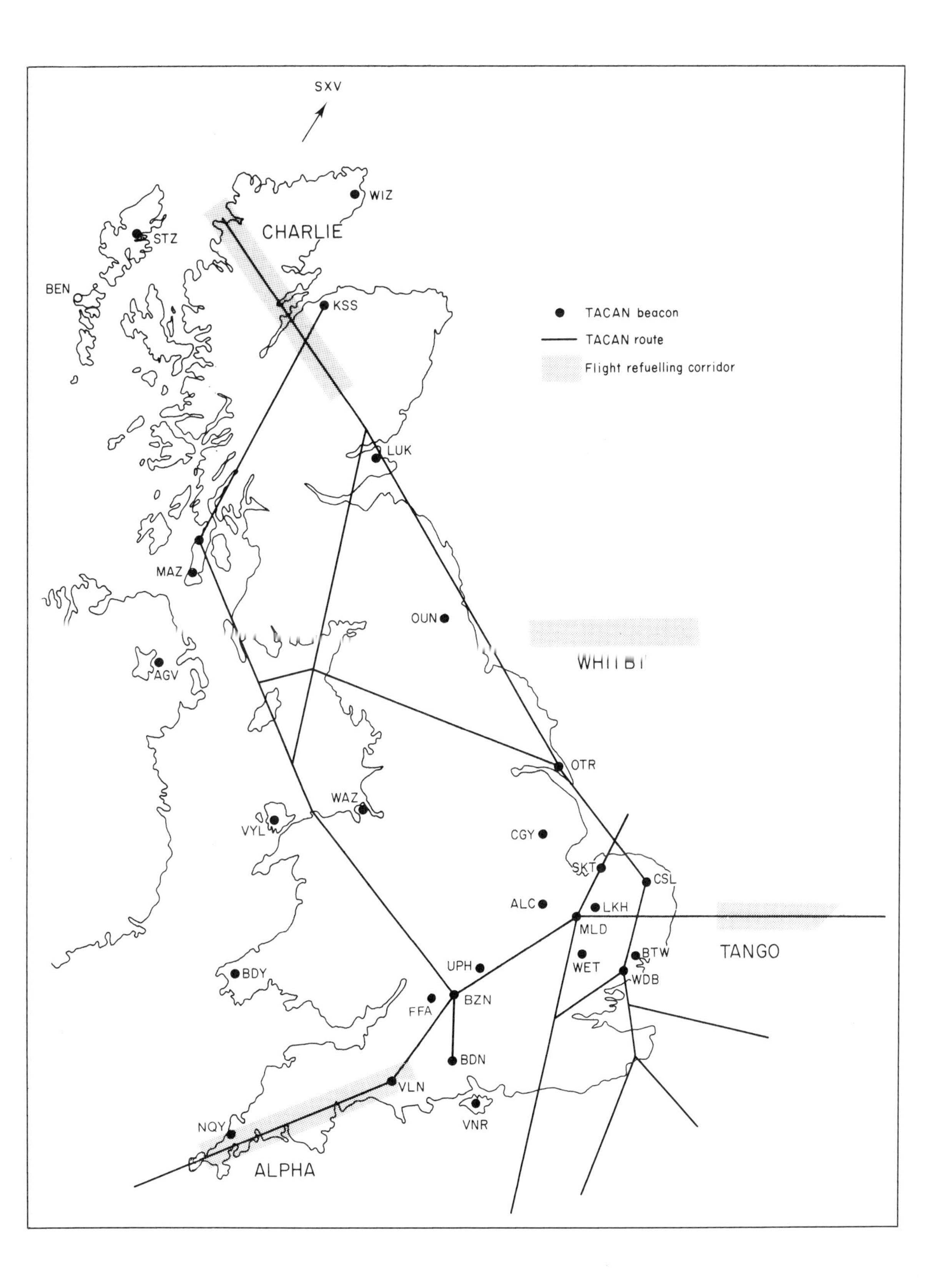

Figure 67 A Harrier receives sustenance from a Victor tanker aircraft. Rendezvous between the two aircraft is made using a direction finder aboard the tanker and air-to-air TACAN. *MOD*

Ground-based Systems

The transmitters used by these systems are not at all like beacons; they are synchronised to transmit together in groups called *chains*, each chain consisting of a *Master* and a number of *Slaves*, usually three. To use such a system the aircraft (or ship) tunes to the master and one of the slaves. Any slight time difference between the signals received from each is then measured electronically. The time lag between the pulses (Loran), or phase difference between the two signals (Decca, Omega), will locate the aircraft on a curved position line which can be found on a specially-produced chart. By tuning another slave, another position line can be worked out. Where both intersect on the chart a fix is obtained.

Loran, Decca and Omega

(Long Range Navigation) chains are located in North America and elsewhere throughout the world. Phase-measurement and pulse techniques are combined in the latest version (Loran – C) which is still in use particularly for ocean flying. *Decca* by contrast uses such short baselines between master and slaves that it only possesses short range capability. It is installed mainly in the British Isles and Europe and is said to be extremely accurate. It can tell when a ship is dragging its anchor! In a fast aircraft Decca is difficult to use without a good deal of expertise, therefore a *Track Plotter* or *Flight Log* has been developed. A pen draws the aircraft's track on specially-printed charts. Some British aircraft such as the BAC 1-11 Series 500 are equipped with Decca Flight Logs.

Compared to Loran and Decca, the other ground based system *Omega* seems to have all the advantages. It operates at very long ranges, covering the globe with a single chain of no more than eight transmitters. These are located in Norway, North Dakota, Liberia, Hawaii, La Reunion, Argentina, Australia and Japan and are operated by the US Navy. Despite the very long baselines, great accuracy is possible because of

Figure 68 Symbols used on the AERAD series of Radio Navigation Charts

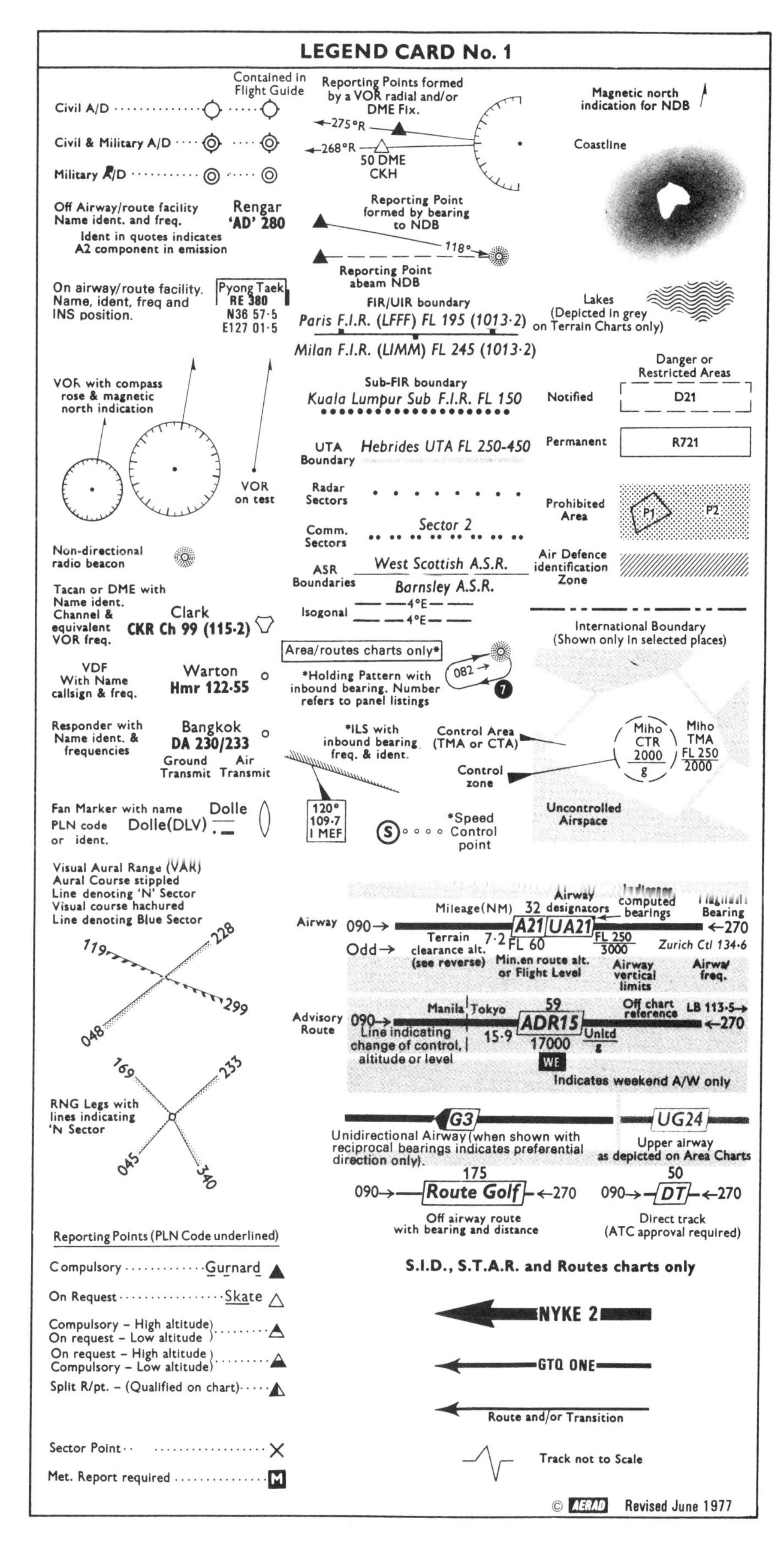

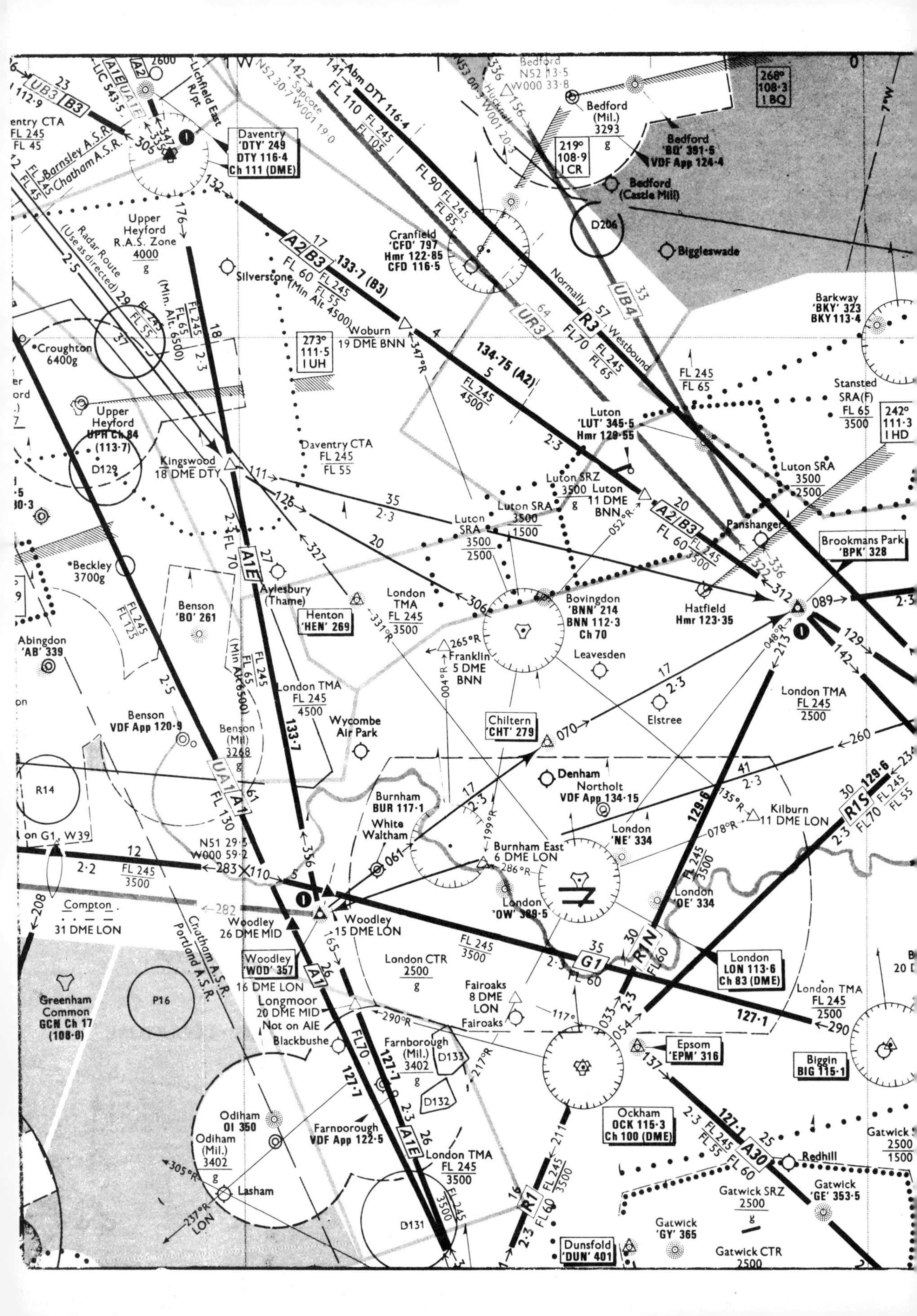

Daventry
'DTY' 249
DTY 116·4
Ch 111 (DME)
Daventry CTA
FL 245
FL 45
Lichfield East
R/pt
Barnsley A.S.R.
Chatham A.S.R.
Abm DTY 116·4
FL 110
FL 245
FL 105
Sapcote
N52 30·7 W001 19·0
Huckna
N53 00·3 W001 20·3
Bedford
N52 13·5
W000 33·8
Bedford
(Mil.)
3293
g
Bedford
'BQ' 391·5
VDF App 124·4
Bedford
(Castle Mill)
268°
108·3
I BQ
219°
108·9
I CR
D206
Biggleswade
7°W
Upper
Heyford
R.A.S. Zone
4000
g
Radar Route
(Use as directed)
Silverstone
133·7 (B3)
FL 60
FL 245
FL 55
(Min Alt 4500)
Cranfield
'CFD' 797
Hmr 122·85
CFD 116·5
FL 90
FL 245
FL 85
Normally
Westbound
FL 70
FL 245
FL 65
Barkway
'BKY' 323
BKY 113·4
*Croughton
6400g
273°
111·5
I UH
Woburn
19 DME BNN
347°R
134·75 (A2)
FL 245
4500
FL 245
FL 65
Stansted
SRA(F)
FL 65
3500
242°
111·3
I HD
Upper
Heyford
UPH Ch 84
(113·7)
D129
(Min. Alt. 6500)
Daventry CTA
FL 245
FL 55
Luton
'LUT' 345·5
Hmr 129·55
Luton SRA
3500
2500
Kingswood
18 DME DTY
Luton SRZ
3500
g
Luton
11 DME
BNN
Luton SRA
3500
1500
Luton
SRA
3500
2500
052°R
Panshanger
Brookmans Park
'BPK' 328
*Beckley
3700g
FL 245
FL 125
Aylesbury
(Thame)
Benson
'BO' 261
Henton
'HEN' 269
London
TMA
FL 245
3500
Bovingdon
'BNN' 214
BNN 112·3
Ch 70
Hatfield
Hmr 123·35
Abingdon
'AB' 339
331°R
265°R
Franklin
5 DME
BNN
Leavesden
048°R
(Min Alt 6500)
London TMA
FL 245
4500
004°R
London TMA
FL 245
2500
Benson
VDF App 120·9
Benson
(Mil)
3268
g
Wycombe
Air Park
Chiltern
'CHT' 279
Elstree
R14
Denham
Northolt
VDF App 134·15
Burnham
BUR 117·1
White
Waltham
199°R
135°R
Kilburn
11 DME LON
on G1, W39
N51 29·3
W000 59·2
Burnham East
6 DME LON
London
'NE' 334
078°R
FL 245
3500
286°R
London
'OW' 389·5
London
'OE' 334
Compton
31 DME LON
Woodley
26 DME MID
Woodley
15 DME LON
Chatham A.S.R.
Portland A.S.R.
Woodley
'WOD' 357
16 DME LON
London CTR
2500
g
FL 245
3500
London
LON 113·6
Ch 83 (DME)
London TMA
FL 245
2500
Greenham
Common
GCN Ch 17
(108·0)
P16
Longmoor
20 DME MID
Not on A1E
Blackbushe
290°R
Fairoaks
8 DME
LON
Fairoaks
117°
Epsom
'EPM' 316
Biggin
BIG 115·1
Farnborough
(Mil.)
3402
g
D133
217°R
D132
Odiham
OI 350
Odiham
(Mil.)
3402
g
Farnborough
VDF App 122·5
Ockham
OCK 115·3
Ch 100 (DME)
Gatwick
2500
1500
Redhill
305°R
237°R
LON
Lasham
London TMA
FL 245
3500
D131
Gatwick SRZ
2500
g
Gatwick
'GE' 353·5
Gatwick
'GY' 365
Dunsfold
'DUN' 401
Gatwick CTR
2500

Figure 69 Reproduction of portion of London Area AERAD Chart, full-size, showing beacons and airways. The original chart is printed in several colours. *Reproduced by courtesy of British Airways*

the unique method of transmission which allows phase comparison not only on the carrier waves but also on their beat frequency. There ought to be special Omega charts showing curved position lines, but these are unnecessary; all the geographical data can be stored in the memory of a micro processor! As a result all that shows on the flight deck is a standard $4\frac{1}{2}'' \times 4\frac{1}{2}''$ black box with an input keypad. When the set is working the latitude and longitude are displayed continually in Light Emitting Diodes. The system may also be coupled to an autopilot which will fly the plane to any chosen waypoint. VLF/Omega, as it is called, is currently one of the most sought-after long-range navigation systems, particularly because of its accuracy and because it is so easy to use.

Radar

Of the *aircraft-based* systems, the first that we shall describe is radar. In World War Two a very successful bombing raid on Leipzig in 1944 was pressed home with the help of high-definition radar which drew a map of the ground below. In those days the system was nicknamed H_2S but was essentially the same as what we now call *Airborne Search Radar* (*ASR*). This operates on short wavelengths – 10 cm. or even 3 cm. – and echoes are returned from quite small objects, including rain. Civil pilots use ASR mostly for detecting and avoiding turbulent cloud formations, whereas military pilots use it for intercepting targets. Pointed downwards, features of the terrain can be reproduced on the radar screen; coastal features, lakes, etc. tend to show clearly.

Doppler

This is a method which uses radio waves bounced back off the ground to measure groundspeed and drift. Speed is measured by utilising the *Doppler Effect* whereby a frequency shift is produced if either the transmitter or reflector are moving relative to each other. A similar device is employed by the police to catch out speeding motorists, and its accuracy is notorious! Used in the air, Doppler is not a complete navigation system but it can be used to improve dead reckoning. A navigational computer employing compass and Doppler inputs is called a *Ground Position Indicator*. It can be used as a bombing aid, or to update moving map displays of the flight-log variety. Such a moving map display, with both compass and Doppler inputs, is fitted to the flight deck of the Trident airliner.

Figure 70 This Omega navigation system will give a readout of an aircraft's position correct to less than a nautical mile anywhere on the globe. *Marconi Avionics Limited*

Inertial Navigation Systems (INS)

Found at one time only on the largest aircraft (747, Concorde, Tristar, etc.). They use super-accurate gyroscopes, of a type called accelerometers, to measure any change in the aircraft's attitude, direction or speed. This information can be computed to keep track of the aircraft's position with an accuracy which in the case of Concorde on a transatlantic run can be better than one mile in 1500. Waypoints can be programmed into an INS, starting with the latitude and longitude of the pier at which the aircraft is parked before the flight. For this purpose the latitude and longitude are nowadays painted on the piers at large airports! Because of INS, Aerad charts now show the latitude and longitude of radio beacons.

Although the inertial navigation system installed in these large aircraft is triplexed, and is a large bulky and complicated package, smaller packages are available which will fit planes of the biz-jet size. Waypoint programming and autopilot coupling is standard.

The military application of INS is *INAS* (*Inertial Navigation Attack System*) which is fitted in most modern strike aircraft, enabling targets to be found at night or in instrument conditions, without having to rely on ground-based aids such as TACAN, which could easily be put out of action in a counter-attack by the enemy.

Calibrating Navaids

If you are a keen airwatcher you will have probably noticed conspiciously-painted aircraft flying low from time-to-time over the countryside and making dummy landing approaches. What you have seen were probably the calibration aircraft used for flight-checking navigation and landing aids. The Civil Aviation Authority operates two HS 748s (G-AVXI and G-AVXJ) and an HS 125 (G-AVDX), painted red and white with black trim and based at Stansted. They are often seen up and down the country checking landing aids at civil airfields. The Ministry of Defence also operate a number of calibration aircraft. 115 Squadron, based at Brize Norton, specialise in this role, using Andovers. Also in use are Canberras and a number of aircraft belonging to the Royal Aircraft Establishment which are engaged normally on experimental work. These are Andovers and 748s.

NDBs do not require calibration; if they should at any time go off the air or alter frequency, the first to know would probably be a special monitoring unit based at the Post Office radio station at Rugby. Perhaps you have often wondered what all those masts were for; this then is part of your explanation! DMEs also present few problems. The navaids which mainly need flight-checking are the VORs and TACANs, especially VORs. Their accuracy depends not only on local features, it can also degrade with the passing of time, so regular checking is essential.

Figure 71 Bristling with aerials, one of the calibration aircraft operated by the Civil Aviation Authority photographed at its Stansted base. This is one of two specially-built and equipped HS748s

VORs are flight checked in two ways. First of all the calibration aircraft flies along the important radials which mark the airways from one beacon to the next to see if the indicated radial is the same as the geographical one. To check all the radials it is necessary to orbit the beacon, typically at a radius of 20 miles at different altitudes. Position has to be checked by any methods available (i.e. Decca), which is accurate enough for the task, and in the UK, where we are well-supplied with Decca chains, this presents no problem. In other places it might be necessary to resort to triangulation from DMEs, aerial photography or, if all else fails, a trained observer stationed at the beacon equipped with a radio who tracks the aircraft visually using a theodolite.

Things to Do

1 Using the suggestions given in this chapter try to find beacons of different types to sketch or photograph. Check that you get the landowner's permission before you trespass. Try to sketch or photograph at least one beacon of each of the following types:
   (a) NDB
   (b) VOR ('dustbin' type)
   (c) VOR/DME ('dustbin' type)
   (d) Doppler VOR
   (e) TACAN
   (f) Marker beacon on ILS approach flightpath.

2 Try to obtain an aeronautical radio chart or RAF En Route Chart on which your local beacons are shown. It may surprise you to find that they are positioned where they are until you start to line them up with each other using a ruler.
   Try to answer the questions:
   (a) Why is this beacon here and not anywhere else?
   (b) With which other beacons does it line up?
   (c) Is there or was there ever an airfield here?
   (d) Any other special local circumstances, e.g. Daventry?

3 If you get a chance to inspect an aircraft flight deck concentrate especially on the navigational instruments. Collect as many photographs as you can of aircraft flight decks and try to identify the various systems.

4 On a clear day study the aircraft flying overhead. Do they change direction when they reach a beacon? Which airways do you think you can see?

5 Find a warm clear day and park yourself near a beacon. You can spend hours here watching aircraft flying overhead, although some of them will be rather high. You can use an air-band radio to identify some of these over-flights, although this is breaking the law (see page 61).

6 If you can manage to get a flight in a small aircraft ask the pilot to show you how he navigates using beacons.

6-Airspace and Air Traffic Control

Like most things in our modern world, Air Traffic Control began in a small way. Controllers worked from the control tower supervising aircraft taking off and landing, moving around on the airfield etc., and in the days before radio managing as best as they could with Verey pistols and hand signals. But as technology developed along came radio, direction finders, beacons and finally radar. Aircraft could now be directed when they were miles from the field, fair weather or foul, approaching to land or en route. Even when they are over the oceans, planes are now 'controlled'. Air Traffic Control now embraces most of the world's airspace which is carved up and parcelled out among the various national ATC authorities. In Britain air traffic control is administered jointly by the Ministry of Defence and the Civil Aviation Authority. These have set up the National Air Traffic Services (NATS) with its nerve centre at West Drayton. This collaboration is important because although the main purpose of ATC is to prevent collisions between aircraft, there is the secondary task of watching out for unlogged intruders entering our airspace. These could be criminal, or perhaps even hostile in intent. The nerve centre of ATC in the UK is the London Air Traffic Control Centre at West Drayton near Uxbridge. This is able to keep watch on the whole of UK airspace, which for administrative reasons is subdivided into the London and the Scottish *FIRs* (*Flight Information Regions*). These FIRs extend from ground level up to FL 245, but air traffic control does not stop there. There are also the London and Scottish *UIRs* (*Upper Information Regions*) extending from FL 245 up to FL 660. However, in the rarified air above 66,000 feet you're on your own. As they say, the only controller on duty up there is God.

Air traffic control today involves complex techniques and equipment, rules and procedures, the full details of which are well beyond the scope of this book. I write with the enthusiastic amateur chiefly in mind, the airwatcher who may find a slight working knowledge of ATC helpful. Should any reader find the treatment sketchy perhaps he will be encouraged to seek elsewhere for more detail.

The Flight Rules System

The person chiefly responsible for flight safety is the pilot, the captain of an aircraft. Many collisions can be avoided merely by keeping a sharp lookout – by seeing and being seen. The *Visual Flight Rules* (*VFR*) are drawn up with this in mind. You keep a good lookout and you hope others are doing the same. If you need to alter course to avoid another aircraft, you turn to the right. If you are following a feature on the ground, such as a coastline or a railway, you should keep to the right. There is no need to file a flightplan (unless going abroad), to keep to airways or to bother with tedious ATC procedures.

However VFR is not possible unless you can remain in good visibility (*VMC – Visual Meteorological Conditions*). You should be well clear of cloud if you wish to fly VFR. The alternative is to fly *IFR* (*Instrument Flight Rules*) which makes two main stipulations:

1 *Terrain Clearance* The pilot should maintain at least 1000 feet obstacle clearance above the highest obstacle within five miles of his track.
2 *Cruising Levels* The pilot should fly the Quadrantal Rule where it applies in the UK. The compass rose is divided into four quadrants NE, SE, SW and NW; the pilot must choose his cruising level from among those allocated to the quadrant in which his magnetic track lies. Thus on a direct flight from Edinburgh to London, a pilot would be tracking approximately 160°. This lies in the south-eastern quadrant so the pilot should select an 'odd' cruising level plus 500 feet, i.e. FL 55. This system tries to ensure that if aircraft are on a collision course at the same level, they may be converging slowly enough to see each other in time.

IFR is designed for *Instrument Meteorological Conditions – IMC*. That is when weather is worse than the VMC limits. But IFR can be flown at any time, even in fine clear weather. IFR *must* be used at night, in upper airspace and at transonic speeds, along the airways and in many control zones and areas at most times. Also the operating rules of many large commer-

Figure 72 Limits of UK airspace, showing boundaries of London and Scottish FIRs

cial aircraft require them to be flown IFR whenever they are carrying goods or passengers 'for hire or reward'.

The Controlled Airspace System

Whereas the simple basic pilot-applied VFR or IFR system may be okay for the less dense traffic routes, along the busy airways and in the proximity of major airfields a better method of keeping aircraft separate and ensuring flight safety is required. That system is the Air Traffic Controller – a ground-based officer who keeps a plot of all the aircraft within his sector. He prevents them from getting too close to each other by using R/T to direct the pilots to manoeuvre accordingly. However before Air Traffic Control of this type can be applied certain pre-conditions are required:

1 There must be sufficient beacons, markers, VORs, DMEs etc., for pilots to be able to report their position and to navigate easily on the controller's instructions. A network of radio and radar ground stations is also necessary, plus a supply of trained controllers, technicians and a host of back-up personnel.
2 Aircraft have to be correspondingly equipped: 2 VHF COMM radios, 2 VHF NAV radios, an ADF, VOR, ILS, DME and a device called a Transponder make up the minimum package for IFR flight under Air Traffic Control. Up-to-date charts and flight guides must also be carried.
3 But it is the pilot who is most affected by the ATC system. He is no longer fully in command of his own ship – he is under orders. He must hold an advanced qualification known as the *Instrument Rating* to prove that he is competent to fly using all this radio equipment, and that he can navigate and report his position accurately. The IR is only granted to pilots who have logged enough hours on their licence and have passed a stiff practical test.

Before the flight the pilot must file a *flightplan* with ATC, giving details of the aircraft, proposed route, cruising speeds and preferred levels, etc. He cannot now take off until he has been cleared to use the airspace he wants. He must keep at least one radio tuned to the controller throughout the flight, and as he flies into another sector he must change frequency as directed to the new controller. He must give the required position reports and estimates, navigate and manoeuvre accurately as prescribed by the controllers.

Extent of Controlled Airspace

However desirable, it would be uneconomical as well as unwise to extend the Controlled Airspace system to cover all of the UK. What about pilots without an IR, private pilots, student pilots, military aircraft on exercises, etc.? In fact the extent of UK airspace which is rigidly controlled in this way is quite small, and as can be seen from the map it consists of the *airways* network – running mostly Northwest Southeast across the nation – and the large *Terminal Areas* around London, Manchester and the Scottish airports. There are also arrangements known as *Special Rules Zones* (*SRZs*) which provide a kind of controlled airspace in the vicinity of some provincial airports: Southend, Luton, Newcastle, Aberdeen, Leeds, Liverpool, etc. The sketch-map (fig. 73) tries to show the layout – but it is difficult to do this accurately with something three-dimensional such as airspace. For example some of the SRZs mentioned above cannot be shown on our map because they are overlain by airways and TMAs, etc. The sketch-map should serve merely as a guide. For more detailed information the reader should consult a radio navigation chart.

Controlled Airspace usually starts at about 3000 feet or above and extends upwards. Thus a private pilot without an IR can fly all over the country provided he keeps below the airways, and dodges round the SRZs, MATZs (Military Aerodrome Traffic Zones), CTRs (Control Zones) etc. which are always depicted on topographical charts. A pilot without an Instrument Rating may still enter controlled airspace for the purpose of landing or taking off provided he files a flightplan requesting a Special VFR flight.

The low-level topographical chart also shows areas which are designated as Danger (D), Prohibited (P) or Restricted (R). These could include nuclear establishments, sensitive military areas (Ulster), wildlife sanctuaries, powerful transmitters, captive balloons and firing ranges. Some of these areas are of special interest to aviation enthusiasts as they are used by the RAF, RN and USAF as bombing, gunnery and rocketry ranges. They will be referred to later in the chapter on military flying (Chapter Eight).

The Controller's Job

When a pilot flies in controlled airspace, he hands over to the controller some of the responsibility for the safety of the aircraft and the conduct of the flight, just as a traindriver must rely partly on signalmen to get his train safely from King's Cross to Edinburgh. Though his job may seem easy at times, an Air Traffic Control Officer is highly trained and must be able to

Figure 73 Airspace arrangements in the UK – Beware! Airspace is three-dimensional; this sketch-map isn't!

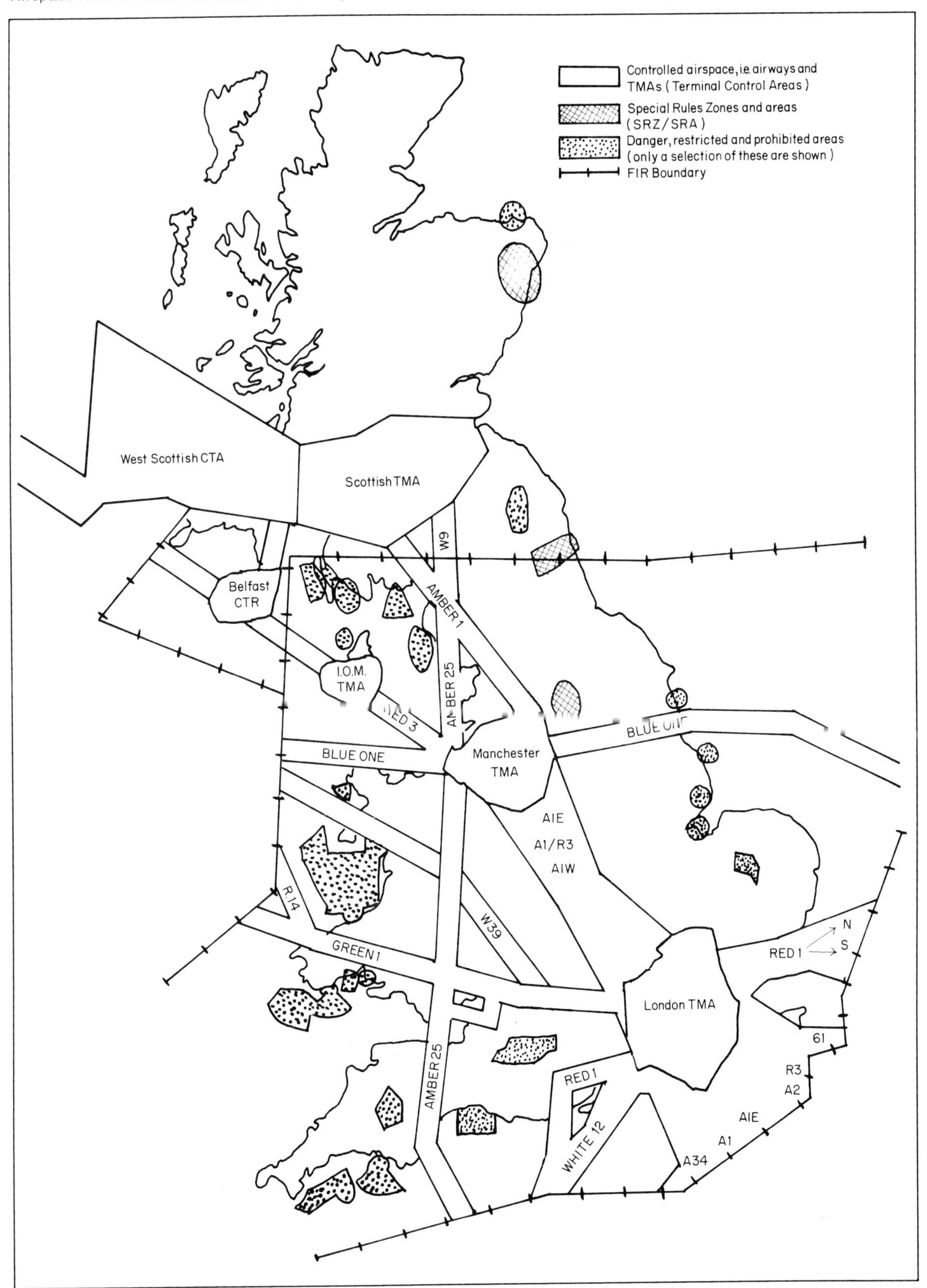

make the correct decision quickly and accurately against a background which is a veritable Hampton Court Maze of rules, regulations and procedures. All this time aircraft may be closing on each other at speeds of over a thousand miles per hour. As pilots say: 'When you see a dot on the windscreen, that's OK. But when that dot sprouts wings, boy, then you're in trouble.' No system of airborne radar so far devised is capable of preventing collisions between civil aircraft. Control has to be from the ground, where the Air Traffic Controller, like a railway signalman, can see the whole picture and can remain in control of it.

In the days before radar, controllers relied on *procedures*. Planes had to fly along definite tracks, conceived rather like railway lines. Pilots reported their position frequently and gave estimates of the time of arrival at their next reporting point. These were usually beacons of the NDB type, later VORs. Over the sea, and where there were no beacons, reporting points had to be estimated by dead reckoning. In this way controllers knew where an aircraft had got to along an airway, and strict *horizontal separation* was maintained. No two aircraft at the same flight level were allowed to get too close. The plane behind could be ordered to reduce speed, or if necessary, to orbit or to hold. Holding involves flying around in the familiar racetrack pattern over a beacon.

Similarly, *vertical separation* is carefully maintained by the controller, and it is up to him rather than to the pilot to decide at what flight levels a plane may fly. Thanks to the sensitivity of modern altimeters, vertical separation of 1000 feet between aircraft is easily attained.

In the old days of procedural control it used to be said that an airliner was 1000 feet deep, 10 miles wide and 100 miles long. This amount of separation might seem excessive, and it definitely restricted the number of aircraft each sector controller could handle at any one time. Nowadays, however, controllers use radar and this enables them to see where each plane is, and the rules regarding separation can be relaxed accordingly.

The Airways Network

Our present system of airways dates back to just after World War Two. The airway from London to Shannon, Green One (G1) was opened in 1950, the first in the elaborate network of airways which today cross the UK. Glancing at an airways map a person might wonder why the main spine of the system runs out of the London TMA towards the Northwest. The reason is because most of the traffic is in transit, not landing in Britain at all, but continuing to or from Europe, Ireland or the US. Stand on a hill in the Lake District on a clear day and watch all the large jets passing up and down Amber One. Most of these planes are from abroad; the UK airways system forms part of the main trunk route between Europe and the US.

Each airway is a corridor of controlled airspace ten miles wide, starting at about FL 55 and extending upwards to FL 245. Above FL 245 is a comparable network of routes called *Upper Airways*. These upper airways are tracks rather than corridors, as all of upper airspace is deemed to be controlled.

The airways are controlled from the ATC Centres at West Drayton, Prestwick and Manchester using radio frequencies in the band from 124 to 136 MHz. An airway is divided up into *sectors* of convenient length, each sector with its own allotted frequency, corresponding in most cases to a separate desk or console in the control centre. However the system is flexible. At busy times the sectors are shortened, more controllers are on duty and more frequencies are in use. At night and other slack periods, sectors hundreds of miles long may be controlled from a single console using the one frequency throughout. In all cases the pilot is told as he takes off which airways frequency to tune, and as his flight progresses, the controllers will tell him when to change to the frequency for the next sector.

As the pilot is directed by the controller to QSY (change frequency) he acknowledges the instructions and reads back the new frequency before switching over. Little is left to chance, because if the pilot should misdial the new frequency, the controllers will have lost contact with him until he spots his error. Nonetheless, details of the system of airways frequencies are available in many publications, including the UKAP and more inexpensive booklets such as the RAF's En Route Supplement. As the airways frequencies are continually subject to change, there is little point in giving many details here.

A pilot may use any combination of airways he wishes to get to his destination, and he will give details of this routeing on his ATC flightplan. However he must expect to be routed via the full quota of beacons and reporting points, keeping to the airways tracks, unless ATC gives him clearance to route direct.

Dotted along each airway is a succession of *Reporting Points*. Some are marked with beacons, but many are not and the positions of these have to be estimated using DME or DR. A pilot will call in to confirm that he has reached a given reporting point and he may also give estimated times of arriving at the next one or two reporting points along his route. Nowadays, the development of radar has made position-reporting much less important than it once was.

Before take-off a pilot will file a *flightplan* with Air Traffic Control. This flightplan tells the controllers everything they need to know about the flight, such as aircraft registration and type, routeing, preferred cruising speed and level, etc. The flightplan is made out in a special code for easy transmission by teleprinter, and even before the flight is airborne each controller who will 'work' the aircraft has the essential details in front of him in the form of a little 25mm. × 200 mm. strip of card which is called a *Flight Progress Strip*. Each card is then mounted in a plastic holder. Meanwhile the pilot waiting at the holding point to enter the runway, hears the controller say: 'Golf Bravo Alpha Oscar Echo is cleared to Prestwick via Pole Hill and Amber One. Squawk 6461.'

Control Using Surveillance Radar

It is quite feasible to control aircraft along an airway as used to be done, merely by shuffling flight strips. Happily nowadays radar provides most air traffic control centres with greater flexibility, enabling the controllers to see aircraft, as their positions change from second-to-second, on the map-like display of the radar screen. Aircraft on converging tracks can be vectored clear of each other and crossing traffic can be slotted-in easily.

At the heart of the system is the familiar rotating radar scanner, or *radar head*. Working on a fixed carrier-wave frequency, it transmits as it scans a succession of short high energy pulses outwards in a narrow beam. At the end of each pulse the equipment is switched to receive and echoes returning from distant objects are picked up. At the control centre these echoes are displayed as blips on the screen. A problem is that not only will ships and aircraft return clear echoes, but so also will trees, hills, buildings,

Figure 74 The London Air Traffic Control Centre, West Drayton, Middlesex. *Civil Aviation Authority*

Figure 75 Anatomy of a Flight Progress Strip used by Air Traffic Controllers

Figure 76 Radar Heads at Burrington, Devon. The scanner in the foreground is used for Secondary Surveillance Radar and is linked to West Drayton. *Civil Aviation Authority*

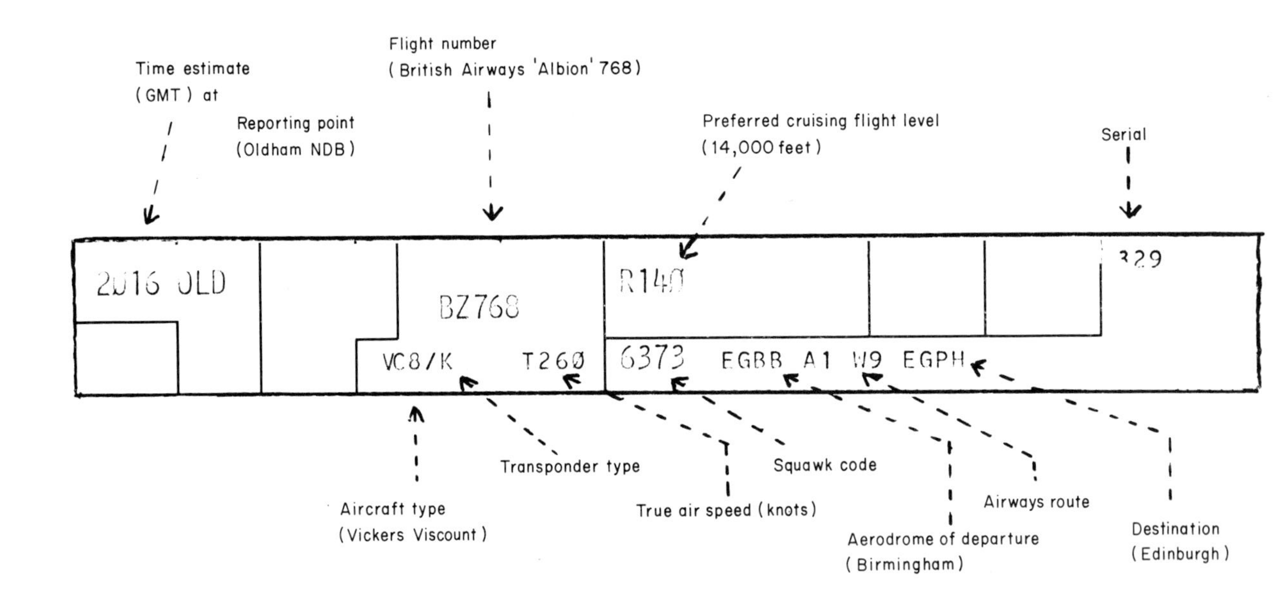

waves at sea, cloud and rain. These can produce permanent or semi-permanent unwanted echoes which are referred to as clutter.

There are a number of ways of getting rid of clutter. Careful choice of carrier wavelength does help, because the longer wavelengths are reflected by weather to a much lesser extent. However 50 cm. radar installations are large and cumbersome. (Visitors to the spectator viewing area on the northern perimeter at Heathrow will have noticed the large 50 cm. Marconi 264 radar head close by).

Electronic methods of eliminating weather clutter, such as variable polarisation and a technique relying on the Doppler effect which filters out all except moving targets (MTI – Moving Target Identification) make it possible to use 10 cm. radars for much Air Traffic Control. These are convenient and compact, such as the Plessey AR 15/2 shown. The vital statistics of this type of radar would be: *Frequency*: A spot frequency in each of either of two bands, 2700–2900 MHz and 3000–3040 MHz. *Pulse Length*: 1 microsecond. *Pulse Recurrence Frequency*: 700 Hz. *Range*: Approx. 80 n.m. max.

On his radar screen the Controller sees a video map of the area with the boundaries of control zones, airways, etc. clearly marked. Aircraft should show as clear blips corresponding to their correct positions. They are returning an echo (but no other data), and one blip looks much like another. With this technique, which is called *Primary Radar*, there is no direct method of identifying which aircraft is which. A blip on the radar screen which could be an aircraft is termed a target.

Most targets will be unidentified until they call in on radio to announce themselves. The controller will ask them their position and will probably ask them to carry out an S-turn which will show on radar. 'Oscar Echo. Will you please turn right now 30° for purposes of identification, then resume your present heading.' The pilot complies. 'Thank you Oscar Echo. You are identified 30 miles south of the airfield.' Many of the targets on the controller's radar may remain unidentified. Unless they themselves call in, the controller has no means of calling them up. He cannot ask them what height they are cruising at so he has no means of separating them vertically from other traffic. If two targets converge and the controller is only in contact with one of them, he will have to ask it to take avoiding action by steering right. The reality may be that one of the aircraft is cruising at 33,000 feet and the other is skulking around at 5000 feet. There was never any risk of a collision but the controller had no means of knowing this. Lack of any kind of height indication is a serious disadvantage of primary radar.

Secondary Surveillance Radar

Nowadays most aircraft flying IFR in controlled airspace are required to carry a little black box called a *Transponder*. Using this device, any four-digit identification code (called a Squawk) can be made to appear beside the aircraft's blip on the controller's radar screen. The device can also be used to encode the aircraft's altitude and this too will be displayed to the controller. 'Oscar Echo. Squawk 6461 and ident.' The pilot acknowledges the instruction from the controller, and dials the code 6461 in the window of his transponder. This squawk appears on the radar screen together with the altitude expressed as a flight level. Also when the pilot presses the little button marked IDENT the blip itself will be highlighted or flash on and off. No need for 30°right turns here! 'Thank you Oscar Echo. You are identified.'

The transponder was developed in the first place as the wartime device known as *IFF – Identification Friend or Foe*. Although civil pilots may refer to the device as an ATC Transponder, military pilots still call it IFF/SSR. The SSR stands for *Secondary Surveillance Radar*. An SSR head transmits on 1030 MHz and receives coded replies from the aircraft's transponder on 1090 MHz. The altered frequency means that only actively transponding aircraft will show on the radar screen – no echoes or clutter. There is a saving of power too; the use of transponders means that the powerful radar emissions needed to produce clear echoes are unnecessary. The squawk number on the transponder is encoded as a set of 12 binary digits. The more mathematically minded readers will know that this allows the pilot to select any one of 4096 transponder codes or squawks. This is best done using an octal scale (one which has no 8 or 9), so the pilot can ring up any code between 0000 and 7777 in his transponder window. The codes are selected by the controller according to a system whereby the digits can tell those in the know something about the origin and status of the flight, who is controlling it and so on. For example a squawk beginning with a 7 is being worked by military radar, 7600 is used in the event of emergency or failure of the communications radio, 5660 denotes a royal helicopter flight and so on. One squawk was allocated to be used in the event of hijack, the transponder quietly informing all controllers of the crew and passengers' plight.

Modern transponders have built into them their own altimeter, a pressure cell which is set to 1013.2 millibars. When the pilot sets the operating switch to ALT or 'Mode C' the transponder can encode the altitude correct to the nearest 100 feet and send this down the line to the radar head. The aircraft's Flight Level is now shown on the radar screen just under-

neath the squawk code. A controller ought to check the altimeter squawk against the pilot's own report of his altitude. If the altimeter squawk is wrong the pilot may be asked to switch this facility off.

Armed with secondary radar, the controller has a formidable tool to assist him in his job. Positive identification is easy and there is a clear and continuous readout of the aircraft's altitude. Provided he can see and think in three dimensions and make decisions in good time, the task of controlling aircraft by keeping them separate either vertically or horizontally looks easy enough. However, secondary radar is not the answer to every problem. Some aircraft have no transponders, and others may not have remembered to switch them on! Controllers who deal with approach zones or blocks of uncontrolled airspace may prefer to use primary radar, because in this way they can see *all* the targets, whether they are squawking or not.

The London ATCC

The nerve-centre for Air Traffic Control in the UK, both civil and military, is the London Air Traffic Control Centre at West Drayton, just off the A4 near Heathrow.

On the civil side, most controlled airspace in the UK is controlled from here, and microwave links to the major airports ensure that the approach and zone controllers at these places (Heathrow, Gatwick, Manchester) can have an up-to-date synthetic radar display. The view inside the Centre, reproduced here, shows dozens of controllers at work, each responsible for some sector of the airways network.

An IBM 9020D computer has recently been installed at West Drayton. This is used, not for controlling aircraft, but for processing flightplans. Flightplans for all scheduled flights into, out of, or over the UK are stored in the computer's memory. One-off flights can be added on a keyboard as the flightplans are filed, and terminals for the same purpose are located at a number of UK airports, each

Figure 77 Air traffic controllers using secondary radar at Manchester. Note the clarity of the radar display and the video detail showing the airways network. *Norman Edwards Associates*

Figure 78 Part of the London Air Traffic Control Centre's IBM 9020D computer installation at West Drayton. *Civil Aviation Authority*

with a direct link to the computer. As the time for take-off approaches the computer allocates each aircraft a squawk code number and starts printing the first of a series of the famous flight strips. Again, the flight strip may be printed as far away as Manchester or Prestwick on a special teleprinter which is linked to the computer. Air Traffic Control Assistants tear off these strips and pass them on to the controller responsible. This is the first part of the *Flight Plan Processing System.*

As many air travellers are well aware, an aircraft's take-off is often delayed for one reason or another. The FPPS will do nothing therefore until the flight gets under way. The computer will know when the flight is airborne, either because the radar network will have started to pick up the squawk, or because a human operator has entered the take-off time. It would take human operators large amounts of time to complete accurately what the computer now does in a matter of seconds as the flight commences. It prints and despatches a complete set of flight strips covering all of the flights within the UK FIRs with time estimates for each sector, and notifies (if necessary) the control authorities of the neighbouring FIRs. Printing flight strips is the computer's main job.

Another facility which the computer has is the *CCCC – Code/Callsign Conversion Cell.* This is able to translate the squawk code originally allocated by the computer back into certain details which it retrieves from the stored flightplan. Thus merely by switching into the computer's data link the controller can erase the squawk code from his radar display, and call up instead from the CCCC the aircraft's registration or flight number plus a two-letter designator which indicates the aircraft's destination aerodrome within the UK, or alternatively the point of exit. All this is now displayed on the screen and alongside it the aircraft's current Flight Level. The wonders of modern technology!

It is of course mandatory to file a flightplan if you are using controlled airspace. It is also mandatory to file a flightplan if you are entering or leaving the

country, even if flying VFR. Thus the London ATCC ought to possess flightplans of all legitimate civil traffic or of military traffic using controlled airspace. The Soviet Tu-114 'Bear-D' shown in figure 79 did not file a flightplan. On radar he showed as an unidentified and unexpected target, which is why a pair of Phantoms (from HMS Ark Royal) were sent to greet him. Large Soviet aircraft such as Bears and Badgers on Elint (Electronic Intelligence) reconnaissance missions frequently invade British airspace. They are probing the NATO defences, just as we probe theirs. At least it serves to keep the UK Air Defence System on its toes. Let us hope that one of these harmless encounters on the bus-run to Cuba doesn't one day turn out to be the real thing. In the meantime there is the ever-present threat of an aggressor slipping through merely by posing as an airliner, or by hiding under the skirts of one. It is worth scrambling fighters to investigate anything suspicious, even if all it turns out to be is a Japan Air Lines 747 which has come over the Pole slightly ahead of its flightplan.

As a result therefore, West Drayton plays host not only to the civil ATC nerve centre, but to a military ATC centre also. Both civil and military authorities work in close harmony under the same roof. Defence is the chief preoccupation of the military controllers at West Drayton, and one way of securing our defence is by checking over the shoulders of the civil controllers that no Unidentified Flying Objects enter our airspace.

Shanwick OACC

In the vintage days of transatlantic flying after the end of World War Two, pilots chose whichever tracks suited them. The shortest distance would be a great circle (which you get by stretching a piece of string between two places on a globe). The great circle route from London to New York sweeps northward, taking in western Ireland and then running down the eastern seaboard of North America by way of Newfoundland and Nova Scotia. A traveller from London to Los Angeles will fly over the cold, inhospitable wastes of the northern Atlantic, Greenland and northern Canada before dropping down into sunny California. Thus on the western seaboard of the British Isles a series of VOR beacons is provided for aircraft setting out over or arriving from the Atlantic: Stornoway, Benbecula, Tiree, Skipness, Macrihanish, Tory Island, Eagle Island, Shannon, Cork and Land's End. These then feed into the airways system.

Great circle tracks would be deviated from where there might be a good chance of picking up strong tailwinds. In upper airspace jet-stream winds of 100 m.p.h. or more can be found, and it pays to get these blowing from behind. A deep depression out to the west of Ireland would have strong winds swirling round it in an anticlockwise manner. Westbound pilots would favour a northerly track, and eastbound pilots would want to swing to the south of this cyclone in order to get the most favourable tailwinds.

Figure 79 An unannounced intruder into British airspace; a TU-114 Bear-D which had been shadowing a NATO exercise is in turn shadowed by a Royal Navy Phantom. *MOD*

Figure 80 'Shanwick, Shanwick, Good Morning. This is Speedbird 861 for Toronto. Requesting Track Bravo and estimating five-nine north, one-zero west at 11.48. Flight Level 330, Mach zero decimal eight zero.' *British Airways*

In a free-for-all situation everybody would want to fly the same track, dangerous nowadays with so many aircraft on the North Atlantic routes. In order to regulate this, the free-for-all of the 1940s and early 50s was ended and an Oceanic Control System was established. The Atlantic west of 30°W is administered by the Gander Oceanic Control Centre. East of here the Shanwick Oceanic Control Centre is responsible. This is based mainly at Prestwick, but there is an important radio link at Shannon – hence the name Shanwick.

Shanwick and Gander between them allocate a number of official tracks crossing the North Atlantic between Newfoundland on one side and the European seaboard on the other, up to ten tracks in all. These tracks are mostly parallel, about 60 n.m. apart, and are delineated each day on the basis of the prevailing weather. The tracks are published in the evening (GMT) and are valid for 24 hours.

The tracks are codenamed Alpha, Bravo, Charlie, etc. with Track Alpha being the most northerly. Depending on conditions some tracks are going to be

Figure 81 Upper Airspace arrangements, showing Upper Airways

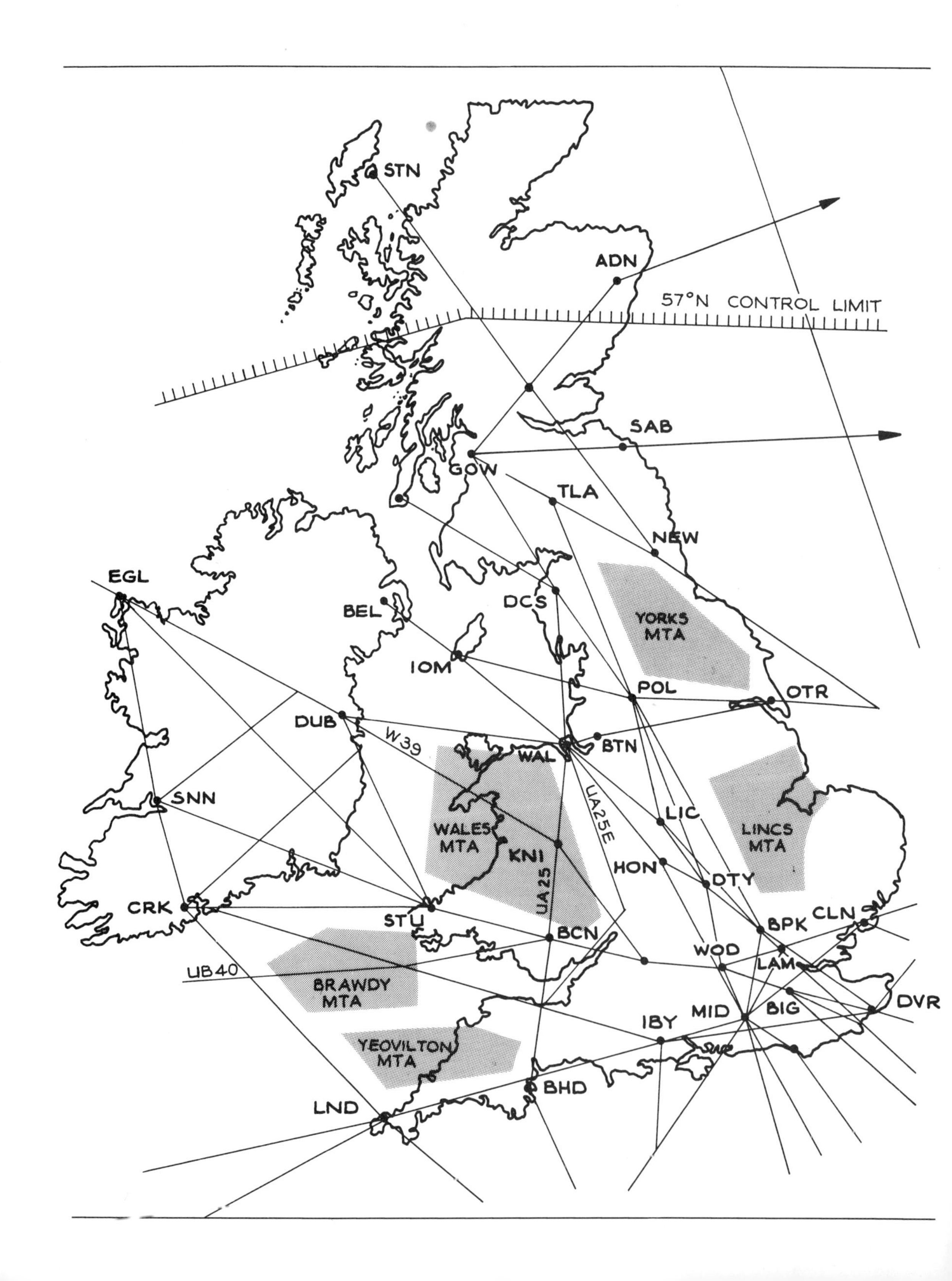

more sought-after than others. It is when long-distance pilots get to somewhere around Pole Hill that they are sufficiently free of local traffic arrangements to tune their VHF radios to 123.95 MHz or 127.65 MHz and to ask Shanwick for the clearance they want. If you live in northern England and you have listened in on these frequencies on an air-band radio, you will have heard much discussion of tracks, time estimates to arrive at the FIR exit points, cruising levels and cruising Mach numbers. Aircraft which are going further afield than the eastern seaboard of North America would probably not need to use the official Track system, and will seek clearance over a succession of waypoints instead. If you listen in to Shanwick, it helps to have a good map of the North Atlantic handy.

Special routes have also been worked out for pilots crossing the Atlantic the hard way – in small- to medium-sized aircraft with refuelling stops in Iceland and Greenland. Many 'Biz-jets' are able to go this way. One stipulation is that HF radio is carried in order to keep in contact with the OACCs at Gander or Shanwick. VHF radio is of no use in mid-Atlantic, so the complex HF communication systems must be used instead.

Upper Airspace

The upper airways follow tracks which are similar to but not quite identical with those of the ordinary airways below. All of upper airspace is under mandatory radar control, except the Scottish UIR north of the 57th parallel and those specially-designated *MTAs – Military Training Areas*. The MTAs are used by military aircraft to practise interceptions, mock combat, etc., and civilian traffic would be vectored clear. A number of upper airways cross the Brawdy and North Wales MTAs and when these are active aircraft wanting to use the airways have to go round the long way. UW39 is available only at weekends. UA25 is also affected midweek, and traffic has to use

Figure 82 Military aircraft use Upper Airspace for much of their training. *British Aerospace*

the UA25E dogleg instead. However when a Concorde flight wants to go supersonic on SWB2, it is the Brawdy MTA traffic which has to keep clear.

Uncontrolled Airspace

Apart from the Quadrantal Rules which govern the choice of cruising level, the only system for preventing collisions which operates outside of controlled airspace is a voluntary one. It is for a pilot himself to decide whether he will call up the Flight Information Service and whether or not to take their advice.

The Flight Information Service (FIS) exists to monitor aircraft movements in the UK FIRs outside of controlled airspace. They depend on the voluntary co-operation of pilots giving position reports and estimates, and abiding by any advice given. They use methods of control by procedure with radar back-up. The FIS operates *Advisory Air Routes* along certain busy tracks. These poor man's airways, as they have been called, exist mostly in northern Britain, e.g. Delta White 2 (DW2) from the Isle of Man to Pole Hill, DG27 (Blackpool-IOM-PWK), DW11 (DUB-IOM-DCS-NEW), DR23 (TLA-SAB-Denmark), DB22 (GOW-ADN-Norway). Many other advisory air routes radiate out from Prestwick and Aberdeen. The radio frequencies for the Flight Information Services are: North of Scotland: 134.0, West Scotland: 124.9, East Scotland: 133.2, Northern England: 134.7, SE England: 124.6, South and SW England: 124.75.

The biggest non-user of the advisory airspace system is the RAF who claim that the FIS uses methods of control unsuited to military aircraft. However the RAF does its best to provide an advisory control system of its own using its radar network. During 'office hours' a pilot outside of controlled airspace can obtain separation service from the many RAF stations which operate good primary radar equipment. The gaps in the coverage provided in this way can be filled by calling the joint civil/military radar network which operates powerful secondary as well as primary radar, such as Border Radar at Boulmer in Northumberland. The RAF is pleased to be able to provide this service to civilian pilots as it makes its own job of ensuring the safe passage of military aircraft that much easier. Another useful advisory radar service is provided in the Thames Valley by the USAF at Upper Heyford.

Things to Do

1 Try to arrange a visit to the control tower at your local airfield. This is easier if you belong to a club or bona-fide aviation society. However if your local airfield is a military one, permission may not be so easy to obtain.

2 When you get to the tower observe the controllers at work. Note what they do and how they do it. If there is radar, the screen may be housed in a separate room. Look for teleprinters and flight-strips.

3 On a clear day observe aircraft using the Upper Airways overhead. Can you determine which airways are being used?

4 Although it is illegal to listen in on air-band radios, you may have a chance to monitor ATC exchanges legally. Listen for position reports and estimates, squawk codes, crossing clearances, etc.

5 Observing flyovers is a time-honoured pastime of air-band radio enthusiasts. They position themselves under an airway on a clear day and tune in to the upper airways ATC frequency for the sector. They claim that with practice, every aircraft flying overhead can be identified.

6 More air-band radio exercises, unfortunately clandestine:

(a) Collecting squawk codes

(b) Listening to Shanwick on 123.95 or 127.65 and plotting waypoints and tracks on a good map of the North Atlantic. Which are the favoured tracks today?

7-Approach and Landing

What goes up must come down, and getting an aircraft back down again on the runway can be one of the most interesting aspects of flying, whether you are a pilot, passenger or a mere onlooker. Landings are surprisingly similar and the pilot's task is very much the same whether it is a Bulldog, Phantom or Boeing 747 that he has strapped to the seat of his pants. He has to handle a large heavy machine which is racing down out of the sky onto a narrow strip of concrete, wheels down, flaps and high-lift devices fully out. Just before this machine flattens itself on the runway the pilot has to bring the nose up to *flare*, or round out the landing. He then flies the next couple of hundred metres a foot or two above the surface hoping to touch down with hardly a bump at all. During the approach the speed was reduced to just about the minimum and the controls have lost much of their effectiveness just as the plane enters the turbulent layer of air near the ground. Quick and positive reactions are needed to battle with the gusts and to line up at the last minute when landing in a crosswind. Park yourself out along the approach one day, near the threshold another day, or some distance along the runway and watch planes landing. It is a source of endless fascination and interest.

Once the wheels touch down securely the pilot finds himself no longer at the controls of an aircraft flying slow, but in the driving seat of a land vehicle travelling rather too fast and in urgent need of being slowed down. There are brakes of course, but in a large or fast aircraft these need to be used with moderation. If they are allowed to overheat a brake fire can result. The usual technique is to use thrust reversers, spoilers – or even a brake parachute – to help get the speed off and to get the weight of the plane firmly on the wheels before using the brakes. On a wet runway there is a risk of aquaplaning, but most modern surfaces are roughened to reduce this hazard.

Before a pilot can land any aircraft he must make an approach, that is, manoeuvre into a position where he is lined up with the runway at the correct altitude, gear down and flaps out and flying at the correct airspeed to complete a landing. This is easy in conditions of good visibility when the pilot can see just where he is and what he is doing. A good method is to fly a partial circuit consisting of a downwind leg, base leg and short final approach. Fast military jets demonstrate this very well when they do a *run and break*, which is a method of getting a formation of up to four aircraft at a time down safely in quick succession. The formation comes in fast at circuit height, then each aircraft turns sharply into a circuit, 'breaking' in different places so as to line up behind one another on the downwind leg.

However, when conditions are made difficult by cloud, haze or darkness, the pilot needs more than his own unaided eye to get down safely. He will need a good approach lighting system, including a set of VASIs, to help him find and line up with the runway, but before he is within sight of these he will need some kind of radio aid to find the airfield and the runway centreline. An instrument approach made with the help of a radio aid such as the ILS is a somewhat different discipline from the visual approaches described above.

Lighting

A gallery of lights resembling motor car foglamps extends for about half-a-mile along the approach to the runway threshold. These approach lights are directional, so they will only be seen by an aircraft which is already on or near the runway centreline. The lights are amber-yellow in colour and will show through haze and thin cloud – even in daylight. This is a useful reassurance to the pilot who may be peering through the murk hoping to find the runway. Very often the approach lighting is the first thing he is able to see, and at night, if the lights are too bright, the pilot can ask the tower to turn down the intensity. Next the pilot will look for the *VASI* (*Visual Approach Slope Indicator*) lights, which are positioned alongside the runway opposite the touchdown zone. These show one set red, the other set white when the approaching aircraft is on the correct approach slope, which for most civil airfields will be 3°.

Instrument Approaches

Aircraft land on instruments not just because of the weather, but because their operating rules may require them to do so. This is as much as anything a testimony to the reliability and precision of modern ILS installations. So even in good weather planes land as a matter of course on the ILS and if there is no haze you can observe instrument approaches easily.

The ILS (Instrument Landing System) will be described in more detail later. A system of radio transmitters produces patterns of radio beams which overlap to form equisignal tracks. One set of beams produces an equisignal along the runway centreline extending back along the approach flightpath for 25 miles or more. Another array of aerials produces an equisignal which extends upwards from the runway at the hallowed angle of three degrees. A plane only has to fly along both equisignals in order to approach on the ILS. It can follow the ILS in and down until it is clear of cloud, then complete the landing visually.

The ILS installations used today are so precise that planes following the line of the equisignals seem to be glued to rails in the sky. This is very obvious if you observe aircraft coming in to land at Heathrow. Every minute or two another plane swoops in, each in exactly the same piece of sky, at the same altitude as the one that went before. This gets a bit trying for the long-suffering residents of Hounslow or Southall, as one aircraft after another whistles overhead. On a clear day if you look up the flightpath you will see perhaps up to four aircraft all lined up on the ILS and coming towards you down the slope. These are being called in by the approach controllers from the 'stacks' at BNN, LAM, OCK or BIG where they may have been flying the holding pattern waiting to land. Assisted by radar the controller spaces each of these out on the ILS approach, about 4 miles apart. Then they come on relentlessly sliding down the three-degree slope.

There is nothing magical about an angle of 3°. It just happens to be a convenient approach angle for many aircraft and both VASI and ILS systems are usually aligned at this angle. The gradient is equivalent to about 1 in 19 or 300 feet for every nautical mile. This is a useful thumb-rule, not only for pilots but also for plane spotters, who can use it to work out the height of an aircraft if they know how far it is from the threshold. From the ground the gradient seems quite gentle, but from the flight deck it may appear quite steep.

In many modern aircraft the autopilot can be used to fly the aircraft down the ILS, locked onto both the localiser and glidepath equisignals. But the pilot still has a few problems. Staying on the correct glidepath may mean adjusting the throttle carefully, especially in a jet where the engines take a while to respond. Wind may present two problems. Landing in a crosswind, a pilot will have to approach 'with drift on' as they say – you can watch this for yourself on a breezy day – and you will see quite large planes flying sideways! Just before touchdown the pilot turns to

Figure 83 A large airliner 'flares' as it sweeps in to land. *Lufthansa*

line up with the runway. Strong winds also mean turbulence near the ground, and you may see aircraft roll and lurch as they fight to stay level. A special type of turbulence is called *wind shear*, caused by buildings and other obstructions slowing down the wind at ground level. To compensate for a possible sudden drop in airspeed, pilots need to approach somewhat faster if wind shear is reported.

Radio Procedures on the Approach

If the flight is VFR the radio procedures are simple. The pilot merely calls in on the approach frequency of his destination airfield when he thinks it convenient to do so. The approach controller will vector the pilot to within sight of the airfield and will then ask him to QSY to the Tower frequency. If the runway is busy the VFR pilot may be asked to orbit, that is, to fly round in circles (30° bank) until cleared to proceed. 'You are Number One' means that a pilot is next in the landing sequence. 'Left Base' means a base leg from which a left turn onto finals is made.

An IFR flight is cleared to the inbound beacon, and before reporting there the pilot will QSY from the Airways to the Approach frequency. It may be necessary to hold; that is, to fly the precise racetrack-shaped holding pattern around the inbound beacon while awaiting further clearance. The Approach Controller (or the Approach Director) will use radar to guide the aircraft onto the ILS, and will direct the pilot to lose height as appropriate, or to check his speed. Once the pilot is established on the ILS he will be instructed to QSY to the Tower, and as he hears the first set of slow bleeps in his headphones he will call in 'Outer Marker!'

On first making contact with his destination airfield a pilot is given the latest weather details, and the all-important QNH and QFE altimeter settings and will be asked to 'copy' – to check these by reading them back. If there is an ATIS broadcast working the pilot acknowledges this by quoting the code-letter at the beginning of the bulletin. 'I have Information India,' for example.

Figure 84 Runway 35 at Dyce Airport, Aberdeen. The high-intensity approach lights can be a reassuring sight to a pilot just breaking through low cloud

Figure 85 The LBA beacon is on the extended centreline of the main runway at Leeds/Bradford airport, and if overflown on a heading of 326° will provide a good approach line to the runway. The full official procedure is as follows: approach the beacon at not less than 4000 feet altitude (i.e. on the QNH), then leave it on a track of 146°. Descend to 3000 feet in about 3 nautical miles, then carry out a 180° procedure turn. Continue descending to cross the beacon again at 1290 feet on the QFE, on the correct track for the threshold. The decision height is 450 feet (QFE). *Reproduced courtesy of RAF No.1 AIDU*

Weather or Not

Although coming down safely through cloud is what instrument approach systems are all about, most of them have to be broken off if at a prescribed height above the ground the pilot still cannot see the runway clearly enough to make a visual landing. The actual *decision height* depends on a number of factors, such as the airfield, local terrain and obstacles, and how precise is the instrument approach system being used. How these decision heights are arrived at is explained below. Once a pilot has got down to his decision height, and if he is still in cloud, he must break off the approach, overshoot and go round again. If conditions do not improve he may have to divert to another airfield.

Weather is therefore the pilot's big question mark. During the flight he will listen to the weather bulletins on LONDON VOLMET (126.6 and 128.6 MHz). His destination may broadcast arrival ATIS information on the local VOR (BNN, BIG, MAY, BTN, GOW and ADN). Runway visibility, wind speed and direction, weather and cloud are featured. The height of the cloudbase is given, and the extent of cloud cover is given in octas (eighths of the sky covered). For example: 'Visibility 1500 metres in rain. Wind variable 5 knots. Cloud two octas at 600 feet, eight octas at 1000 feet.' This tells the pilot that he will be in thick cloud until he gets below 1000 feet, and there will still be patchy cloud down to about 600 feet. Most published decision heights are below this level, so there should be no problem in completing the landing.

The set of decision heights promulgated for the various runways and systems at a given airfield are known as *Instrument Approach Minima* (*IAM*). They are shown on the various official landing charts for the airfield. These minimum heights are worked out taking into account any tall obstacles not only on the flightpath, but also within reasonable radius. The size of the radius is wide or narrow depending on the precision of the approach aid being used. Thus for an NDB let-down, obstacles within a wide radius would be included. In hilly country the Obstacle Clearance Limit could be as high as 500 feet, and a further 100-foot error margin on top of this would give a decision height of 600 feet. By contrast, the very precise Category II ILS installed at many major airports permits decision heights as low as 170 feet. However, many airlines require their pilots to adopt more stringent IAM limits than those published. 'Check your company minima.'

Non-precision Instrument Approaches

These are fairly simple systems using ground-based direction finders, or beacons such as NDBs.

Using a direction finder such as a VDF, the pilot has to call up the tower for a QDM, in fact a series of such bearings. He homes in to the field until his QDM has become the reciprocal, then continues overflying the field to and fro as he loses height to break out of cloud. Decision height has to be fairly high using this imprecise system, but it does work. At some airfields, particularly military ones, the controllers use their direction finder as though it were a kind of radar, to 'talk the pilot down'. This method is termed a QGH – controlled descent through cloud.

An aircraft which carries its own ADF (Automatic Direction Finder) can let down using a beacon near the airfield. Some beacons are on the extended runway centreline, others are not. The beacons used are NDBs, which are simple, requiring no calibration and little maintenance. Before using a beacon the pilot should have the published *Instrument Approach Chart* for the facility. This will show him the position of the beacon in relation to the airfield and its runways, the appropriate decision heights, spot heights at different points along the approach, plus of course the frequency and callsign. We publish one of these charts here. On the Leeds and Bradford chart the NDB is on the extended runway centreline (see figure 85). If you study these charts you will see how a pilot might use them to get down out of cloud and reasonably lined up with the runway. However you will notice on them at least two strange features – the Holding Pattern and the Procedure Turn. Both of these are an important feature of instrument flying.

The Holding Pattern

If an aircraft needs to hold, it must do so in a prescribed manner. The holding pattern is racetrack-shaped with semicircular ends and straight sides. Each of the straight sides and each of the semicircular ends should be flown for one minute respectively, so that the entire pattern takes 4 minutes. The pattern must be flown precisely as indicated, turning left or right as prescribed and always approaching the beacon along the correct QDM. The turns at the ends are Rate One.

Although the holding pattern is tedious, it is a manoeuvre pilots need to master, as they need to use it quite often. Aircraft approaching Heathrow or other busy airports may be required to hold at peak periods. They fly the holding pattern at different flight levels above and below each other. This is known as *stacking* and the four stacks used by Heathrow Approach are at

CIVIL PROCEDURE Published by 1 AIDU (RAF) **Changes: New Format**

LEEDS APP **123.75**	LEEDS TWR **120.3**		LEEDS RADAR **121.05**
SAFE ALTITUDE	100nm 5500ft	25nm 4100ft	10nm 3200ft

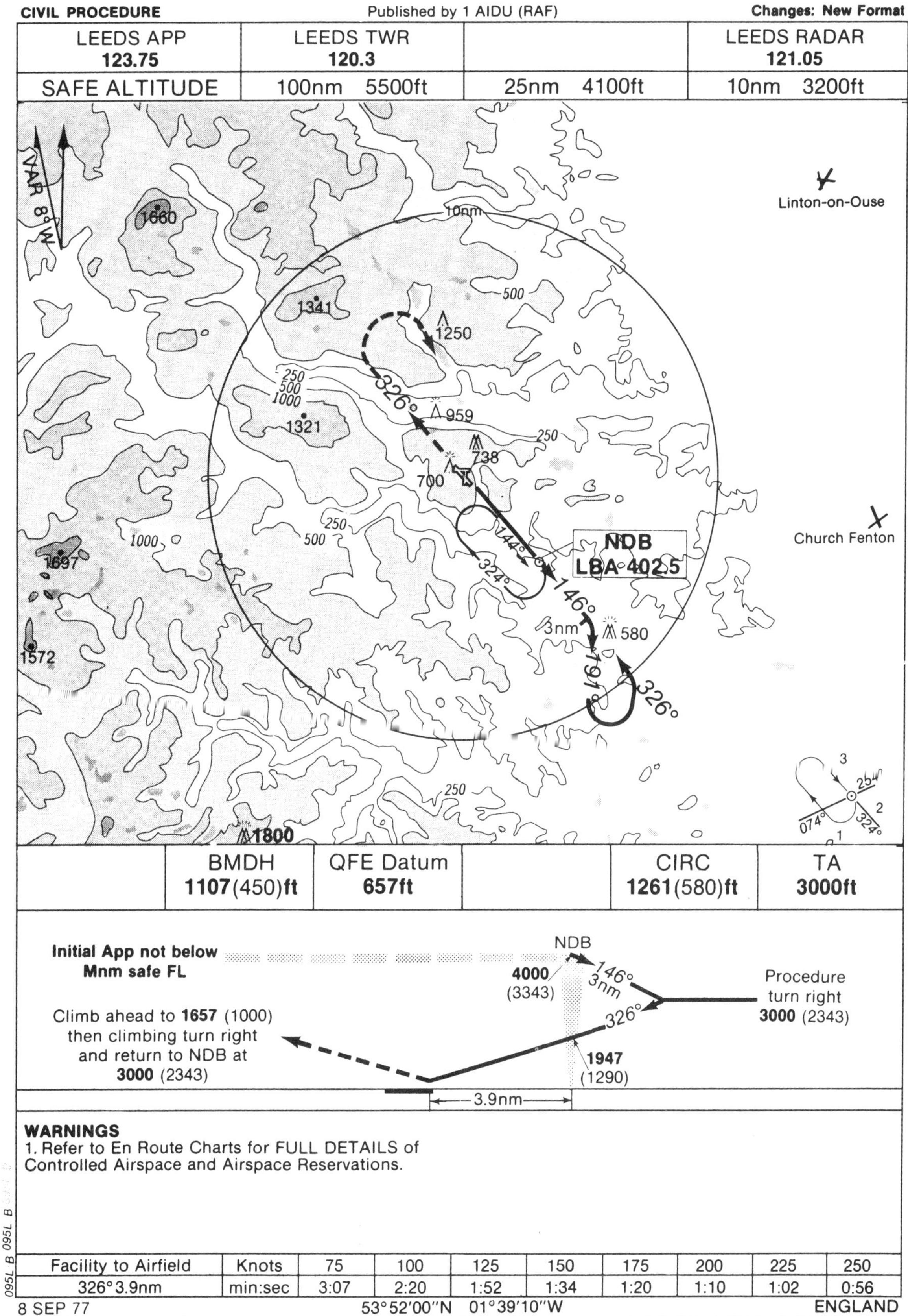

	BMDH	QFE Datum		CIRC	TA
	1107(450)ft	**657ft**		**1261**(580)ft	**3000ft**

Initial App not below Mnm safe FL

NDB

4000 (3343)

146° 3nm

326°

Procedure turn right **3000** (2343)

Climb ahead to **1657** (1000) then climbing turn right and return to NDB at **3000** (2343)

1947 (1290)

3.9nm

WARNINGS

1. Refer to En Route Charts for FULL DETAILS of Controlled Airspace and Airspace Reservations.

Facility to Airfield	Knots	75	100	125	150	175	200	225	250
326° 3.9nm	min:sec	3:07	2:20	1:52	1:34	1:20	1:10	1:02	0:56

095L B 095L B

8 SEP 77 53°52′00″N 01°39′10″W ENGLAND

NDB Rwy 33 **Elev 681ft** **LEEDS/BRADFORD**

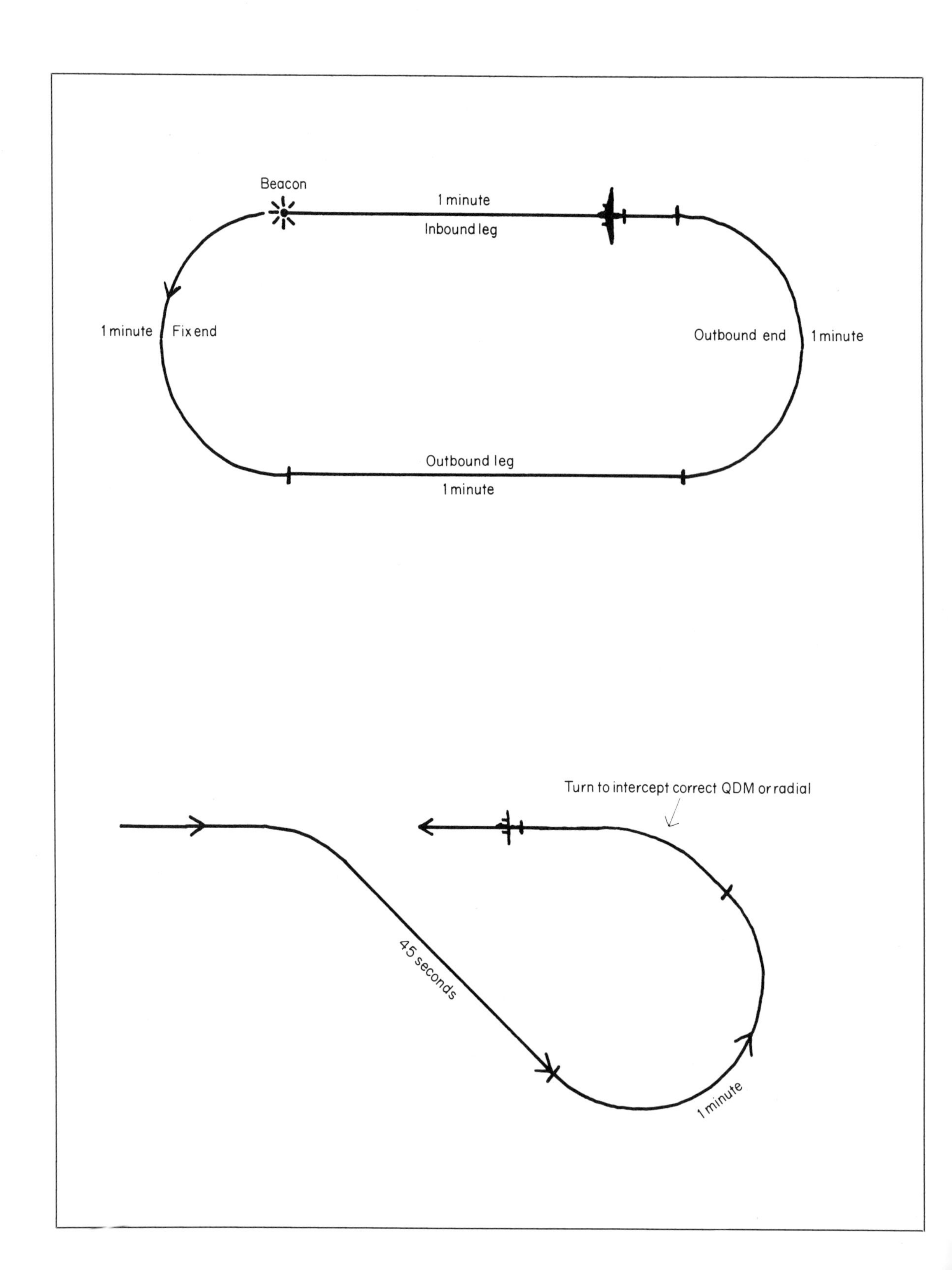

Figure 86 The Holding Pattern (*top*) and the Procedure Turn

Bovingdon (BNN), Lambourne (LAM), Ockham (OCK) and Biggin (BIG). On a clear day you may be able to see the stacks over these beacons. The lowest plane will be at FL 60 or FL 70, and as aircraft are called in from the bottom of the stack the others move down a notch. From time to time you may observe the holding pattern being flown at other beacons.

The Procedure Turn

This is another manoeuvre you may be able to observe. It is a method of carrying out a 180° turn so that you return along your original track already lined up with the let-down beacon and the runway centre-line. You will see the procedure turn precisely depicted on instrument approach charts, and you may be lucky at times to be able to observe the manoeuvre itself.

TACAN and VOR/DME Let-downs

These beacons can be used as let-down aids also. The distance readout makes it easy for the pilot to estimate his position and to judge the correct approach slope. With TACAN, offsets from the beacon will make it easy for quite complex procedures to be flown anywhere in the approach environment.

Your Own Airfield

The chart reproduced in figure 85 is from the series of RAF Terminal Approach Procedures Charts Series. These are available from No. 1 AIDU RAF Northolt, West End Road, Ruislip HA4 6NG, at the very modest price of a few pence each. We recommend that you study approach charts of your local airfield, as the best way of understanding instrument approaches. As with all aeronautical charts, you may be able to obtain secondhand copies, which are ideal.

ILS – Instrument Landing System

The present-day VHF/UHF landing system known as ILS has been in use since just after World War Two, although many of the principles and techniques which ILS embodies were developed before and during the conflict.

From the early A–N radio ranges came the idea of a radio beam taking the form of an equisignal. Using the pre-war Lorenz or Standard Beam Approach system the pilot would listen for the continuous tone which marked the correct flightpath. During the war this system was developed by the Germans into a series of bombing aids with varying success, but they did

Figure 87 An ILS (Instrument Landing System) Localiser transmitter installed at Norwich Airport. *Plessey Radar Limited*

Figure 88 An ILS Glidepath antenna alongside the runway at Guernsey Airport. *Plessey Radar Limited*
Figure 89 ILS Monitor receiver close to the A45 at Birmingham Airport

produce one very useful improvement. Instead of having to listen through headsets to find the equisignal, the pilot merely had to keep a pointer central on a dial. Any deflection of the pointer to left or right told the pilot which way to steer to regain the equisignal. This development is the direct ancestor of today's VOR/ILS meter.

Early landing aids featured a line of beacons along the approach. At each of these the pilot would check his height. NDBs were first used, and some of these *Locator* beacons are still around here and there. Fan markers also came to be used, and today these are more or less standard, except for some newer installations which use DME instead.

A full ILS installation features a number of different bits and pieces scattered around the airfield. First there is the *Localiser* transmitter which produces an equisignal beam along the runway centreline and out along the approach for 25 miles or more. The localiser is a characteristic fence-type structure at the far end of the runway. Situated near the touchdown zone is a lattice tower supporting the aerial array of the *Glidepath* transmitter. This produces another equisignal beam, tilting upwards from the runway at 3°. If a pilot follows both of these beams he can make a very precise instrument approach down to decision height. The *Markers* – ordinary 75 MHz fan marker beacons – are positioned out along the approach, the Outer Marker about 4 miles out and the Middle Marker about one mile out. And finally, to complete the installation it is necessary to position a number of *monitor* receivers here and there around the approach.

Most busy airfields, military as well as civil, have at least one full ILS installation. This will be aligned along the preferred approach to the main runway. Large airfields may have more than one ILS and Heathrow has five.

Meanwhile aboard the aircraft the pilot tunes the

G-BCPF
MIC
CLIMB
VHF 1
OFF
VHF 2
SPKR
OFF REC ADF
TUNE
PULL VOL FOR BFO
TEST
DME 190
MILES MIN KNOTS
114.10
DME
DIM PUSH TEST
IDENT
WING PROP
LANDING GEAR
POSITION

Figure 90 The ILS and VOR meters (to the centre-left of the panel) and the similarity between them is clearly shown in this photograph. The vertical pointer of either meter will respond to an ILS localiser beam or a selected VOR radial. The horizontal pointer of the ILS meter indicates the ILS glidepath

Figure 91 The Instrument Landing System (ILS)

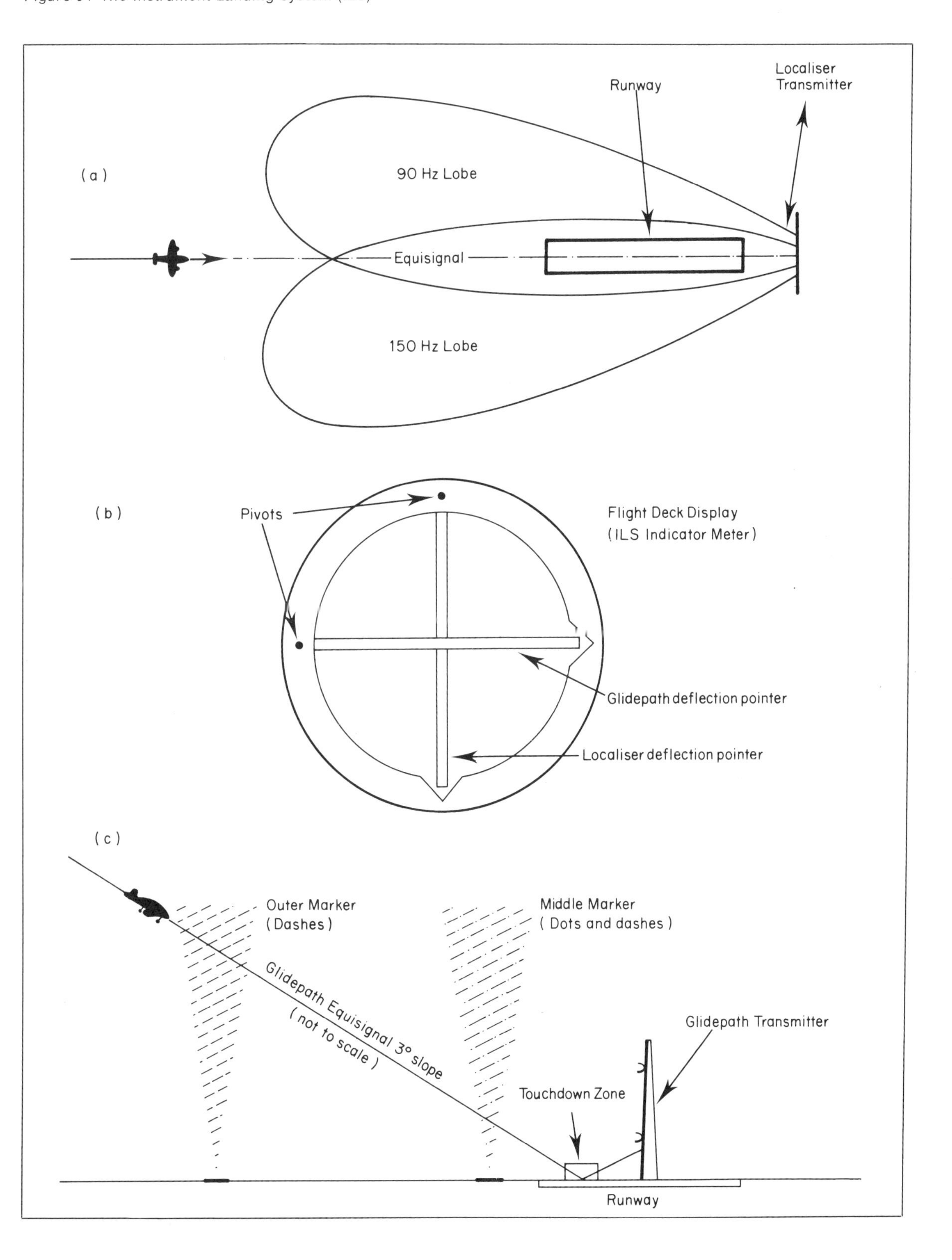

Figure 92 'Oscar Echo, Turn now for right base and steer 130° to intercept the localiser from the right. Call when established.' *British Airways*

frequency of the ILS on his VHF NAV radio and the pointers on his VOR/ILS meter should come alive. ILS was designed to be compatible with VOR, so that an ILS localiser could activate the vertical pointer on a VOR meter. Conversely, one can fly VOR on the vertical pointer of an ILS meter. Think of an ILS localiser as a VOR with only one radial. If the aircraft is lined up on the beam the pointer should centralise, and it will show a deflection if the aircraft is off the beam to either side. A typical ILS equisignal beam is really quite narrow – no more than a few yards wide.

There is another pointer on the ILS meter – a horizontal one. This deflects in exactly the same way to show whether the aircraft is on, above or below the correct 3° glidepath, so to correct you fly up or down as indicated. Flying the ILS looks quite simple: keep both pointers crossed in the middle of the instrument and you can't go far wrong.

Sophisticated aircraft can make an ILS approach using an autopilot which will lock onto the equisignals. You may be surprised that there is no ILS meter as such on these aircraft. The localiser signals are fed into the HSI and the glidescope signals go to the ADI.

The ILS Localiser

This is the conspicuous yellow-painted fence which faces the end of the runway. It comprises an array of dipoles end-on to each other and backed by a reflector fence. It produces an equisignal by transmitting two narrow overlapping lobes, each modulated in a different way. The lobes are in balance and produce an equisignal along the centre of the overlap zone. This is aligned precisely along the runway centreline and out along the approach flightpath for 25 miles or more.

Localisers operate on VHF frequencies – the band from 108 to 112 MHz which they share with VORs. If you have an air-band radio, you can tune in to the localiser and listen to its Morse ident. However you

Figure 93 Outer Marker and NDB. The VHF folded dipoles on the right are the Outer Marker for the ILS approach to Runway 24 at Manchester. Alongside it is the NDB 'Mike Echo' which pilots find useful when lining up on the flightpath. Mike Echo can also be used for emergency holding or as a let-down aid in its own right

will also hear a characteristic buzzing noise which seems to vary depending on where you are standing in relation to the runway. This is the modulation applied to the signal, and if you are standing in the left-hand lobe (as seen from the approach) you should hear mostly the 90 Hz modulation, but if you cross over to the other side of the flightpath you will hear the 150 Hz modulation which is applied to the right-hand lobe. Along the flightpath itself the strength (amplitude) of both modulations should be in balance. It is unlikely that you could detect this with your untrained ear, but the specially-designed ILS meter fitted to an aircraft can compare both lobes easily. And as we have said before, a meter which will work with both VOR and ILS is a useful added bonus.

Dotted around the airfield you will find a number of *ILS monitor* receivers, looking very much like VHF TV aerials. These are also designed to compare the relative strength of both lobes, and if the balance between them changes they switch off the transmitter automatically.

The ILS Glidepath

This transmitter also produces an equisignal by mixing 90 MHz and 150 MHz lobes. The aerials are mounted on a lattice tower near the runway touchdown zone, in such a way that the radiation pattern is reflected off the ground to form an equisignal which slopes upwards at 3°. The radio frequency used for the glidepath is in the UHF band, but this is coded to the VHF localiser frequency in accordance with a pairing system. All the pilot needs to do is to select the ILS localiser frequency on a VHF NAV radio. A special UHF tuner will select the correct glidepath frequency automatically. If for some reason this arrangement does not work (or the glidepath transmitter is U/S) a NO SIGNAL flag will show on the ILS meter.

ILS Marker Beacons

It is usual to instal two beacons of the Fan Marker type (see page 77) along the approach flightpath. These operate on 75 MHz and beam a directional signal upwards in a kind of cone or fan. As a plane overflies these markers the Marker Receiver comes alive and produces the characteristic bleeping noise and winking coloured lights associated with each type of marker.

The *Outer Marker* is located about 4 n.m. from the threshold. The exact distance will be found on your local ILS chart. It emits 2 dashes per second at 400 Hz audio tone and causes a *blue* light on the instrument panel to blink on and off.

Figure 94 Standard Terminal Arrival Chart – London/Heathrow. *Reproduced by permission of the Civil Aviation Authority*

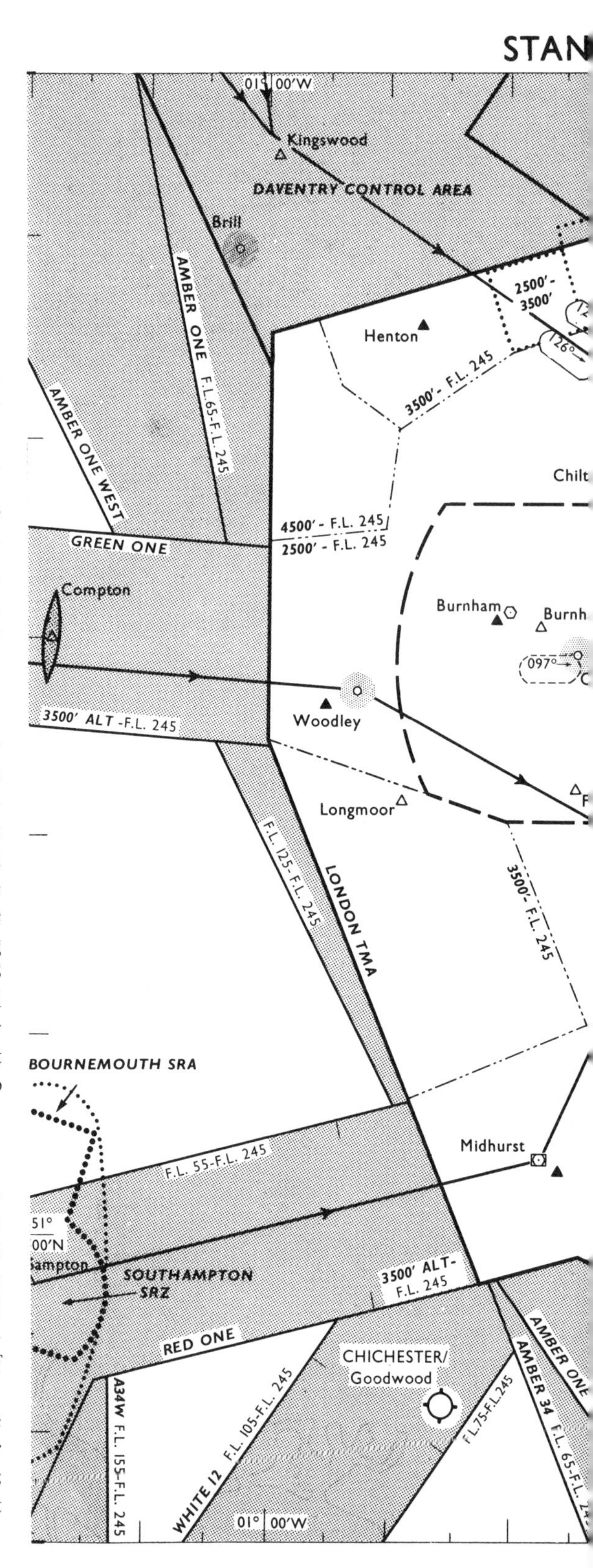

ERMINAL ARRIVAL CHART - LONDON/Heathrow
00° 00′
00° 30′ E
01° 00′ E
SFC-
F.L. 65
Stansted
Luton
2500′-
3500′
Stansted
3500′-ALT-F.L.245
RED ONE NORTH
F.L. 55 - F.L. 245
3500′ - F.L. 65
STANSTED SRZ
SFC-2500′
SFC-3500′
TON SRZ
Braintree
Clacton
Matching
1500′-2500′
Brookmans Park
Tripod
SOUTHEND SRZ
SFC-6500′
Lambourne
270°
RED ONE SOUTH
CTR SFC-2500′
Southend
Kilburn
229°
Hornchurch
O.E.
277°
BLUE 3 F.L. 55 - F.L. 245
MANSTON SRZ
Bromley
Pilgrim
Biggin
311°
Epsom
329°
Manston
3500′-F.L. 245
Detling
1500′-2500′
Sevenoaks
Tonbridge
Marden
Dover
G.E.
265°
085°
G.Y.
LONDON/
Gatwick
WORTHING CONTROL AREA
SFC-2500′ ALT
GATWICK CTR SRZ
Sandgate
Mayfield
310°
305°
Lydd
Battle
Rye
Lydd
245
245
Miller
Plowman
LYDD SRZ
SFC-4500′
Farmer
Cliff
AIE F.L. 55 - F.L. 245
00° 00′
Seaford
01° 00′ E

The *Middle Marker* is less than a mile out. It produces 1300 Hz dots and dashes and activates an *amber* light on the panel.

ILS and DME
Some very new installations use DME instead of marker beacons. The DME facility is positioned alongside the runway half-way point and is callsign- and frequency-paired to the ILS. Distance to go to the runway can be read off continually. Edinburgh's Runway 07/25, Gatwick's 08/26 and Heathrow's parallel runways all have DME.

A Typical ILS Approach
London's Heathrow airport may be the busiest international airport in the world, but there is nothing untypical about the procedures for landing there. At peak periods a pilot may have to spend some time flying the holding pattern in the stack over one of the four inbound beacons. These are Bovingdon (BNN) near Hemel Hempstead, Lambourne (LAM) near Chigwell, Ockham (OCK) near Woking and Biggin (BIG). Each plane will be separated by 1000 feet from the next one below and the lowest aircraft of all will be holding at FL 60 or 70. Departing aircraft also overfly these beacons, but in the slot between 3000 feet (QNH) and FL 60.

As aircraft are siphoned out of the lower levels of the stack, the others move down. Finally it is our pilot's turn to leave. He is already tuned to the Approach Controller (119.2 MHz) and he will be given a series of headings to fly as the Controller monitors the approach on radar. The pilot must control his

Figure 95 A HS 748 calibration aircraft of the CAA makes a low-level pass over a prototype Microwave Landing System installation. The technician is operating a telecroscope. Note the spotlight on the aircraft's nose. *MOD*

speed and reduce his height in accordance with the Controller's instructions. Finally this *radar vectoring*, as it is called, will position the aircraft where it can easily pick up the ILS.

The ILS beam is so narrow that if an aircraft did not close with it at a shallow angle, it would shoot straight through it. Thus it is usual to close with a localiser at an angle of 40° or less, and to pick up the glidepath by flying level at about 2000 feet and intercepting it from below. Once an aircraft is *fully established* on the ILS it is usual for the pilot to QSY to the Tower frequency 118.7 MHz.

It is also customary for pilots to report when they are passing the Outer Marker. They then receive final instructions for landing, including the latest wind data. By now the plane will be less than 1000 feet high and the pilot will be watching his decision height coming up. Suddenly he is out of the cloud and the runway approach lights and VASIs are dead ahead. He punches the AUTOPILOT DISENGAGE button and gets ready to flare the landing.

A Case of the Bends

ILS cannot be installed at every airport. The radiation lobes from the localiser, for example, can be reflected from hills, buildings, and in some cases, vehicles, causing a thickening in part of the overlap area. This will produce either a 'bend' in the equisignal, resulting in a 'bent' flightpath, or worse still a 'multipath' effect caused by freak equisignals. Thus ILS is impossible to instal at airports in mountain valleys or on some rocky Mediterranean islands.

Even where conditions are good and where the ILS is known to perform well, as at most UK airports, the absence of bends and multipaths cannot be taken for granted. The little monitor receivers dotted around should indicate any change in the signal detectable at ground level. However it is the beams in the air which really matter, and the only way to check these is to flight-test the ILS on a regular basis as a matter of routine. As mentioned in Chapter Five in the section on calibrating navaids, both the RAF and the CAA operate specially equipped ILS calibration aircraft.

Some bends can be detected easily. Others will only reveal themselves to aerial photography or to a special device called a *telecroscope*. This is positioned alongside the touchdown zone of the runway and focussed on a spotlamp fitted in the nose of the calibration aircraft as it flies down the ILS approach. The infrared-sensitive telecroscope can track this spotlamp and pinpoint any deviation from a straight flightpath.

Category II ILS

The Instrument Landing Systems being installed today are much more reliable, stable and precise than their predecessors, thanks to technical improvements which have been developed in recent years, such as solid state circuitry, improved aerial arrays, better methods of calibration, etc. At the same time the airborne part of the system has been improved and flightdeck procedures have been tightened up. When all these improvements are taken together we have what is termed *Category II ILS* which can be certificated for operating down to much lower minima, i.e. lower decision heights are possible when the full Cat II specification can be met. For example the ILS on Runway 24 at Manchester specifies an OCL of 176 feet and RVR of 600 metres for Category I operation; however, it is also certificated for Category II operation, permitting an OCL of 116 feet and runway visual range as low as 350 metres. The advantage of Category II capability is that it should reduce the number of occasions when flights have to be diverted because of bad weather – an important consideration for airlines, and for their passengers who dislike the inconvenience which these diversions cause.

Autoland and Category III

Fully automatic landing systems have been developed which make it possible to follow the ILS all the way down to the runway. The final stages of the approach are controlled with the help of radio altimeters and computers, and these *Autoland* systems can flare the landing and fly the plane into the gentlest of touchdowns without any help from the pilot's eyes or hands. These systems were designed to make it possible to land an aircraft in thick fog and they have been fitted to a number of modern aircraft types, including the Trident, VC 10, Concorde and all the wide-bodied jets. Diversion ought to be a thing of the past, but unfortunately it is a phenomenon which is still with us. The problem is one of certification, and before Category III operations are permitted (zero decision height and zero RVR) a large number of technical preconditions must be fulfilled. In practice these conditions cannot always be met, hence aircraft fitted with Autoland are sometimes diverted too.

GCA – Ground Controlled Approach

The controller can talk a pilot down if he has a good enough radar. Most ATC radars are of the surveillance type and although they can be used to help a pilot find the airfield, they lack the precision necessary to deal with really low cloud. Talkdown using ordinary

Figure 96 ILS chart for Manchester's Runway 24. *Reproduced by permission of the Civil Aviation Authority*

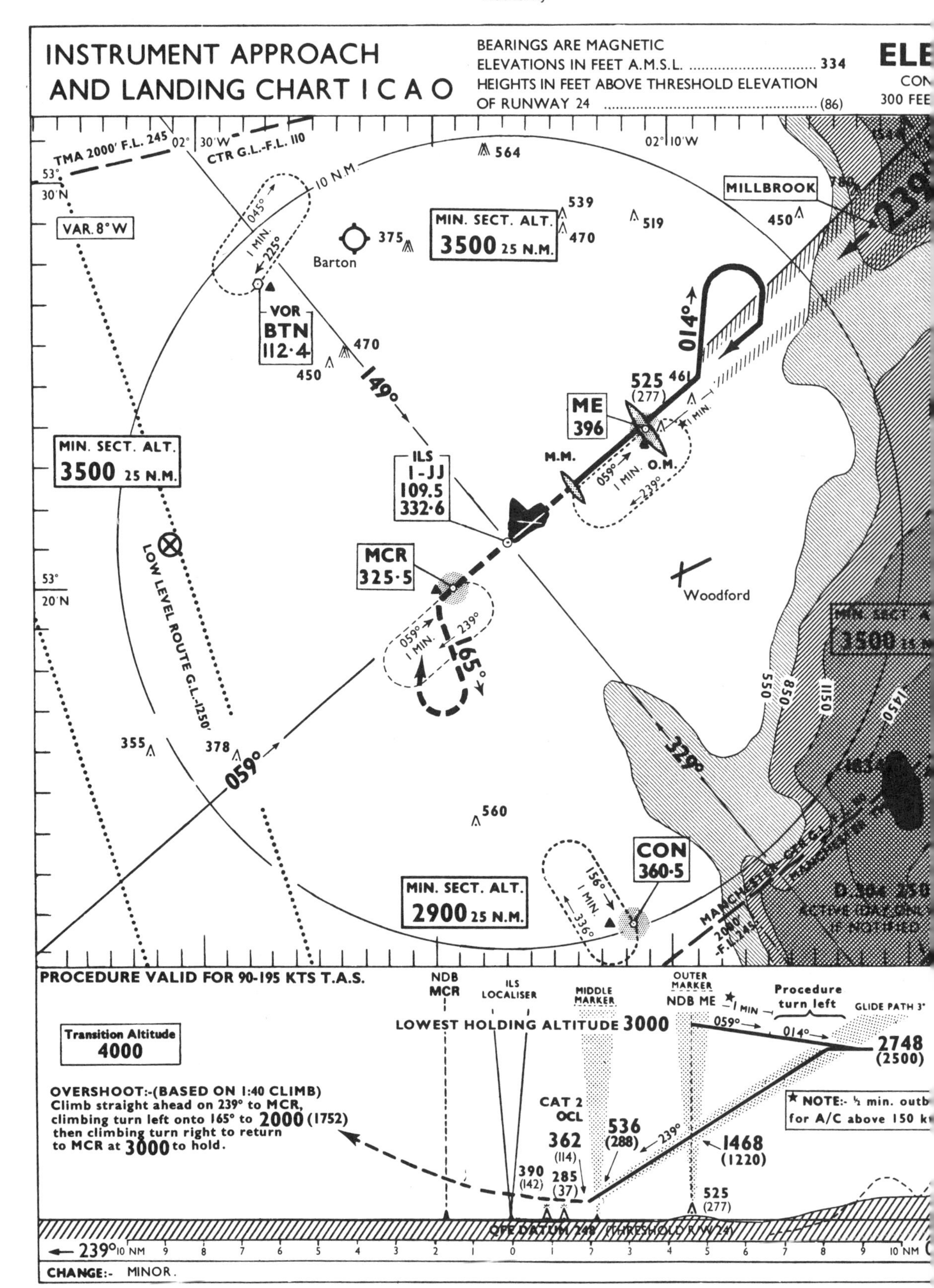

FT. MANDATORY PROCEDURE MANCHESTER/International

ICE ROME CATEGORY II ILS R/W 24

OME ING A	TAKE OFF	DAY	NIGHT
	LANDING	DAY	NIGHT

STANCE BETWEEN MARKERS	90 KTS	120 KTS	140 KTS	160 KTS	180 KTS	195 KTS
to M M 2.92 nms	1 min 57 sec	1 min 28 sec	1 min 15 sec	1 min 6 sec	58 sec	54 sec
to threshold 0.73nms	29 sec	22 sec	19 sec	16 sec	14 sec	13 sec

RADIO

	119·4	MANCHESTER APPROACH
	119·4	MANCHESTER APPROACH
DAR	121·35 119·4	MANCHESTER DIRECTOR
rgency)	121·5	
R	118·7	MANCHESTER TOWER
rgency)	121·5	
C	121·7	MANCHESTER GROUND

LIGHTING

G

4 914M. H.I. coded C.L./ 5 bars. V A S I 's

LTG

4 H.I. and L.I. components with H.I. colour coded C.L. and red end Lights.
Green H.I. threshold with elev. H.I. green wingbars.
900M. T.D.Z. lights.

R LTG

axiways. Obstructions

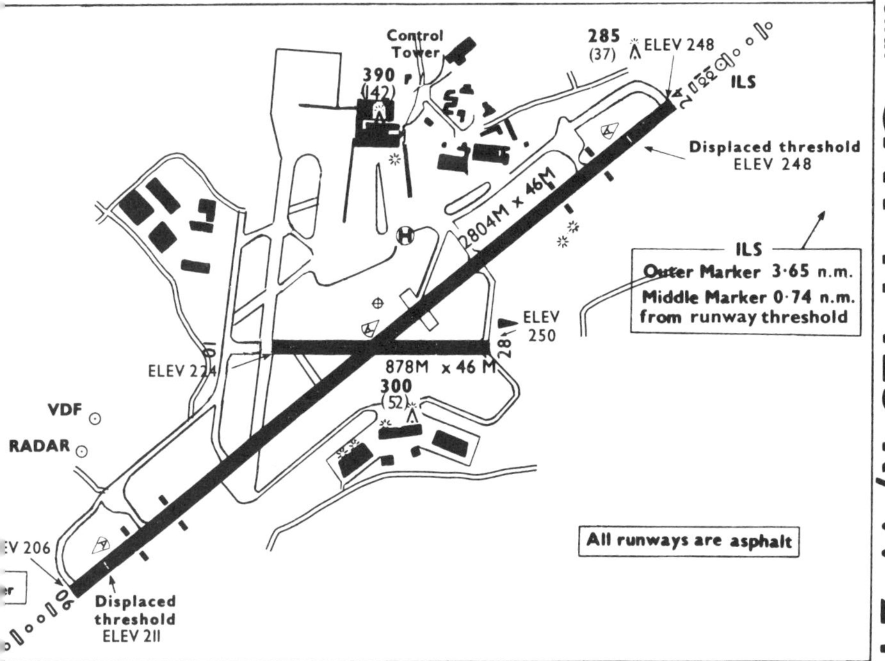

53°21'N 02°16'W MANCHESTER/International CAT II ILS R/W 24

Figure 97 ILS (*top illustration*) and Microwave Landing Systems compared

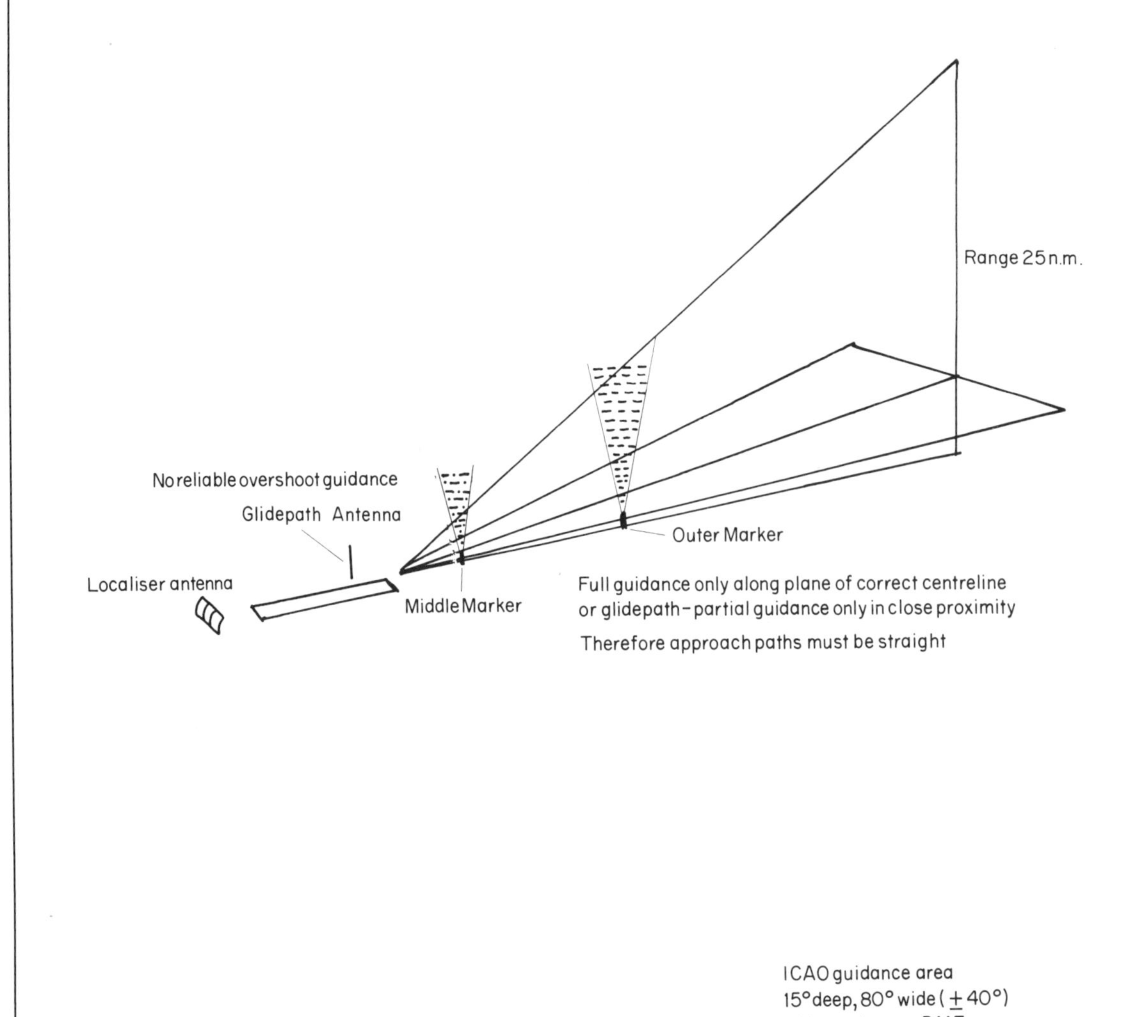

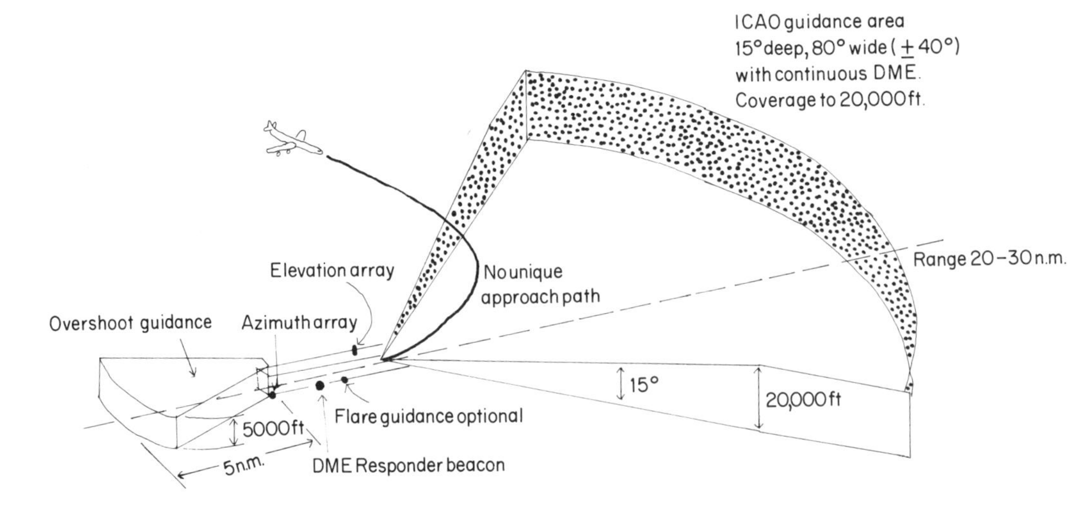

surveillance radar is termed *SRA – Surveillance Radar Approach.*

Several notches better is a *Precision Approach Radar (PAR)* talkdown. You probably will not find PAR at many civil airfields, but the RAF is very fond of it. A little house, shaped somewhat like a funny-looking chapel, can be rotated on a turntable to face along the duty approach. The 'chapel' houses two radar heads, one scanning in azimuth and the other in elevation. Simultaneously the controller watches both displays. A pilot requiring a PAR talkdown tunes to the special PAR radio frequency, which can only accommodate one aircraft at a time. He is vectored to a point about 5 miles from touchdown and a height of about 1500 feet. The talkdown can commence from here and while it is in progress he is asked by the controller not to acknowledge any further instructions.

'Five miles to run. You should be passing 1500 feet. Turn right 5 degrees to regain the centreline. Turn left 5 degrees. On the centreline. On the centreline. A little below the glideslope now. 4 miles from touchdown, your height should be 1200 feet. On the glideslope. Turn left 5 degrees. . . .' The system is a good one as far as safety is concerned. It permits quite low decision heights – 200 feet or less – and because the controller can see the aircraft all the time on radar, the protagonists of PAR maintain that it is much better than ILS; no fear of a pilot flying himself into the ground. However, against this the system is limited by being able to handle only one aircraft or at most two at a time. For this reason ILS is preferred to PAR at busy civil airports.

Figure 98 Precision Approach Radar (PAR) installation; RAF Coningsby

Figure 99 Microwave Landing System. A prototype MLS azimuth array temporarily installed at Heathrow compared for size and appearance with its present-day counterpart – the ILS localiser on Runway 10R. *MOD*

Microwave Landing Systems

Recently (April 1978) an ICAO panel agreed to develop a microwave landing system (MLS) to replace ILS towards the end of the century. A microwave system would be different in many ways from the current VHF/UHF based ILS:

1 The use of frequencies as high as 5 GHz would overcome the problem of 'bends' in the flightpath caused by reflection from hills and buildings. Landing systems could be installed in mountain valleys and at other airfields where ILS had proved difficult in the past. The cost of installation and maintenance would be lower than ILS, and there would be no need for rigorous flight-checking.

2 New features could be incorporated which are not possible with ILS. Overshoot guidance and flare guidance could be provided, and by using TDM (Time Division Multiplexing) other information could be provided, such as weather, or continuous RVR.

3 The system would provide full and accurate guidance over a wide approach angle, and over a range of approach slopes. This would make it possible to fly fully instrument-controlled curved approaches, instead of the present straight-line ILS approach. By varying the approach flightpath, the noise problem could be spread over a wider area. Range-measurement would be by using DME.

The microwave landing system adopted by ICAO will use the TRSB (Time Reference Scanning Beam) principle to provide guidance both in elevation and azimuth. This has been compared to a car's windscreen wiper. A beam scans to and fro and the time difference between the to and fro scan can be measured in the aircraft and made to indicate its position on the approach. Aircraft will need to be fitted with special receivers, decoders, etc., but there is no immediate hurry. MLS is still in the development stage. It is unlikely to be installed at many airfields before 1985.

Things to Do

1 Try to obtain instrument landing charts of your local airfield and study them in conjunction with Ordnance Survey maps of the same area.

2 Try to locate the beacons and the Outer and Middle ILS Markers.

3 If you live or work under a part of the sky where holding patterns or procedure turns are flown, keep an eye out for these.

4 Observe aircraft during an approach. Look out for the following:
  (a) Lining up with the ILS localiser
  (b) Jet engine power bursts during the approach
  (c) The flare
  (d) Approaching with drift on in a crosswind and straightening out to touch down
  (e) The sequence of braking actions, e.g. spoilers, wheels on the runway, reverse thrust, brakes, etc.

5 If you have an air-band radio you will probably want to use it to listen to aircraft all the way down the frequencies. However this would be illegal. Pity, because you would be able to plot the approach on a map or chart.

8-Military Flying

Aviation enthusiasts tend to fall into two camps, those who prefer civil and those who prefer military aircraft. The differences between the two types of aircraft are fairly obvious, but the resemblances are there too. The main flight controls and instruments are the same, but the task which military aircraft are called upon to perform is usually very different from that of their civilian counterparts.

The military air forces stationed in Britain are part of the NATO alliance and are committed to the defence of the West. Hopefully they may be able to achieve this without firing a shot in anger, but it would be foolish to act on this assumption. It would be equally foolish to try to forecast the precise opening moves in any future conflict, whether with the Warsaw Pact forces or some other possible enemy. NATO must prepare for any future threat using whatever strategy from whatever quarter, and must learn to respond accordingly. This is the doctrine of the *Flexible Response* and if the airwatcher thinks about it, he will appreciate much of what he sees happening in and around military airfields. The planes he sees landing and taking off may be visitors from another base and squadron, or even from another NATO country.

Viewing Military Aircraft

Military bases are not open to the public except on special occasions such as Open Days, Battle of Britain displays, etc. However this need not deter the keen enthusiast. If you explore the perimeter of the airfield you should be able to find places where you can get quite good sightings of aircraft parked or manoeuvring on the ground, approaching to land or taking off. The perimeter fence is usually the boundary of Ministry of Defence land, but take care just the same that you do not trespass on private property either. One further point: don't leave any food remnants around which might attract birds. If you're observing military aircraft in any other country (or *any* aircraft in Eastern European countries) check and then double-check that you're not breaking any rules.

Weapon Firing Ranges are places where aircraft practise attacks using live ammunition, smoke bombs, rockets, etc., and if you can get near to them they are good venues for watching military aircraft in action at close quarters. The Pembrey marshes in Dyfed are used by the Hawks from the Tactical Weapons Unit at RAF Brawdy. Other ranges are Jurby off the Isle of Man, West Freugh near Stranraer and Wainfleet in Lincolnshire. When the range is in use a red flag will be flying, and it is advisable not to approach too close.

Some RAF airfields are designated as *Master Diversion Airfields* capable of recovering aircraft of any type which have diverted from elsewhere. Most diversions are caused by bad weather but a small number are the result of equipment failure. These airfields have good round-the-clock crash-landing facilities, such as nylon arrester nets, or *RHAG* (*Rotary Hydraulic Arrester Gear*). This latter piece of equipment is similar to the arrester gear used by the Navy on carrier flight decks. Fast jets are fitted with arrester hooks which they use to engage the gear.

Types of Airfield

On the accompanying map (fig. 102) we have shown the main bases used by the various air arms stationed in the UK: RAF, RN, USAF. The Royal Navy is mainly concerned nowadays with helicopters, as are the Royal Marines and the Army. The MOD (PE) (Ministry of Defence Procurement Executive) is concerned mainly with the testing and evaluating of new aircraft, missiles and equipment.

Much of the service pilot's time is spent training, whatever the nature of his Squadron's duties, but the RAF operates a system of Flying Training Schools which take a pilot from his basic training in a Jet Provost (Linton-on-Ouse or Cranwell) through Advanced Flying Training on Hawks at Valley or Jetstreams at Finningley. Navigators train on Dominies at Finningley and the Tactical Weapons Unit at Brawdy takes over where Valley left off. Helicopter pilots train at Odiham or Shawbury, and Air Traffic Controllers at Shawbury.

R/T Frequencies and Procedures

Military aircraft use frequencies in or near the UHF band for R/T communication. A combined VHF/UHF crystal controlled transceiver is normally fitted so that VHF channels can be tuned when it is necessary to contact civil ATC for airways clearance or when landing at a civil airfield. The callsigns 'Rafair', 'Ascot', MAC and SAM are among those used by military aircraft when calling in on VHF.

Nobody manufactures a receiver which is commercially available for listening in on military UHF, so if you take your air-band to a military airfield, you will be restricted to the few odd bits of civilian traffic which are using the airfield's VHF frequencies.

With military aircraft R/T, security needs to be borne in mind. Aircraft are seldom referred to by type, and the coded callsigns which each aircraft uses are changed periodically. This is to prevent those enthusiastic air-spotters who form the crews of East European trawlers or the staffs of Embassies from

Figure 100 A Breguet Atlantique of the French Aeronavale calls in at a UK base. *Keith Price*

building up too accurate a picture of NATO readiness by listening in on UHF!

The emergency frequency is 243 MHz. This is monitored by ground stations and sometimes by other aircraft in flight. On the ground there is throughout the UK a network of sophisticated direction finders tuned to 243 MHz and linked to the London Military ATCC at West Drayton. An aircraft in distress merely has to press the switch that brings in the emergency transmitter for its position to be pinpointed instantly on a huge wall-panel illuminated by strobe lights. This is quicker than using radar, where there might be delay in determining which of several targets was in difficulties. For the rest, military R/T procedures closely resemble those of civil pilots.

Air Traffic Control

The main function of air traffic control is to maintain safe separation between aircraft. En route control of military aircraft uses a system of seven Military Radar Service Areas covering the country. There are no military airways as such, apart from the TACAN routes, therefore procedural control without assistance from radar would be difficult. In practice there is a chain of secondary radar heads, some of them covering specific areas, such as Border Radar at Boulmer, Highland Radar at Buchan, and Ulster Radar at Bishop's Court.

Most airfields have their own radars. These are usually clutter-free primary outfits such as the Plessey AR 15, although the USAF has secondary radar at all of its airfields. Many RAF controllers who do not have secondary radar are divided in their opinions as to its usefulness. In crowded British airspace, where there is much light aircraft traffic, it pays to be able to see every target, whether it is transponding or not, and for this reason many would want to hang on to their primary radar sets, which they use for surveillance of

Figure 101 A Phantom is brought to an abrupt halt by the arrester gear aboard HMS Ark Royal, the last of the carriers. Similar RHAG (Rotary Hydraulic Arrester Gear) is installed on land at RAF and USAF bases. Tucked away in the foreground, with wings folded, is a turbo-prop Gannet. *MOD – RN*

Figure 102 Chief UK military airfields

the approach control area. Civil aircraft passing through can call in on VHF and ask for an advisory radar service, which the RAF is happy to provide. It takes two aircraft to have a collision and one of them might be their own!

At most military airfields also you will see something I mentioned in the previous chapter, a peculiar building containing the scanners for the Precision Approach Radar or PAR. At military airfields this is the favoured instrument landing aid. There is usually an ILS, which pilots can use if they wish, but they are invited to opt for PAR. In fact it is said that RAF controllers feel snubbed if a pilot refuses a PAR talkdown!

Most civil airfields have now got rid of their PARs, whereas by contrast, many RAF airfields would prefer to rip out their ILS! There are subtle reasons for these different preferences. At busy periods, most civil controllers cannot spare an extra pair of hands to give a PAR talkdown to every Tom, Dick and Harry who calls up wanting one; moreover, operating rules usually specify ILS anyway and it can be locked into the autopilot, whereas PAR cannot. By contrast the RAF is trained to recover each and every one of its aircraft, even if the plane is damaged, the crews tired or battle-weary. In such conditions, a radar talkdown is felt to be the surest way of getting a pilot home safely when the chips are down. The reassuring sound of the controller's voice has a psychological value.

Navigation Techniques

There is a wide variety of navigation systems fitted to military aircraft, depending on the role for which they were designed, or when they were commissioned. Navigation systems employing ground beacons or transmitters, easily 'taken out' by an

Figure 103 A Royal Navy Wasp helicopter on a TOTAL gas rig in the Frigg Field. The helicopter was from the 2200-ton Leander class frigate HMS Charydbis (F75), seen in the background, while on Fishery Protection duties. *RN*

Figure 104 RAF Hawks on the ground. *Keith Price*

enemy, such as TACAN, ADF or Decca, are luxuries which might not be available in wartime.

TACAN is the standard en-route navigation system fitted to almost all military aircraft. A TACAN beacon is found at many airfields, and others have at least an NDB. The ATC procedures based on these aids are exactly the same as the civil counterparts, i.e. the holding pattern and 180° procedure turns. There are some spectacular additional techniques, such as the VFR landing run-in and break, or the *TACAN dive-arc* recovery method. An arc of sky near the airfield and defined by TACAN offsets is set aside for dives through several thousand feet. At about 3000 feet an aircraft can be vectored by radar onto an instrument approach. Sometimes the TACAN is itself used as a landing aid, especially at USAF bases.

Decca was at one time installed in a number of RAF aircraft, and was even intended to be used as a blind-bombing system. However, a better idea, independent of vulnerable ground transmitters, was a *Ground Position Indicator* fed with both compass and Doppler inputs. Both these methods have been superseded by versions of Inertial Navigation systems. INS was described in an earlier chapter. It uses super-accurate gyroscopes and is completely independent of ground stations. Specialised versions of INS have been developed for military aircraft, including a blind bombing technique known as *INAS – Inertial Navigation Attack System*.

Many military aircraft nowadays carry a crew of two – a pilot and a navigator. The navigator is really a full-time radar operator, concerned almost entirely with the aircraft's airborne search radar, which is a military version of the weather radar fitted to civil planes, only larger and more versatile. The scanner is housed in the aircraft's black nosecone, and can be controlled by the navigator using a 'joystick'. Other aircraft will show on the radarscope, so it can be used for interception. Angled down towards the ground it becomes a 'mapping radar' and can be used for very

precise navigating in IMC, and even as a super-accurate bombing aid, if the navigator is experienced enough.

Low-Level Flying

Low-level flying is an important ingredient of modern air warfare, as demonstrated in the Middle East and in Vietnam. It is necessary to fly as low as possible in order to avoid detection by enemy radar. In wartime many sorties would be flown almost entirely at low-level and NATO pilots must have adequate training in low-level flying.

It is not economically feasible to send all of the RAF's Strike Command pilots abroad to Canada or Australia on a regular basis for all of their low-level training. For units based in Britain (and this includes the USAF) there is no alternative but to organise low-level training sorties in the UK itself. However exciting this prospect might be for plane spotters, it can excite strong reactions in other members of the public. Some people are perturbed by the sudden roar of jets overhead at low-level, and many feel that the RAF picks on them for this kind of exercise far too often.

A number of points must be noted. First of all, there is no indiscriminate low-level flying. *All* low-level sorties are planned meticulously beforehand and this discipline applies to all military aircraft. 'Low-level' refers to any sortie below 2000 feet, but it is quite common to get down to as low as 250 feet above the terrain.

A low-level run is planned by the pilot himself using topographical charts or OS maps. The pilot's instructions require him to avoid danger areas, controlled airspace, nuclear establishments, certain petrochemical complexes, etc. In order to reduce disturbance to the absolute minimum the main conurbations are avoided also. A system exists for notifying the Ministry of Defence of any activities such as crop-spraying (or even hang-gliding) which are imcompat-

Figure 105 RAF Hawks in the air. *Keith Price*
Figure 106 Wrap-around camouflage on a Hawk from the Tactical Weapons Unit, RAF Brawdy in a near-vertical bank off the Dyfed coast. *MOD*

ible with low-level flying. The pilot then draws a felt-pen line on his map threading a careful route past all of these areas. The curves on the line are those which the pilot will actually turn, drawn with special stencils, e.g. 400 knots, 30° bank etc. The pilot then has his flightplan cleared by Operations and booked out. He then cuts his low-level run from the map and fits it behind a transparent panel on the thigh of his flying suit. No unplanned low-level sorties are permitted; all must be booked-out beforehand.

The illustration (fig. 110) shows the flightplan for a simulated attack on an electricity pylon. Permission to 'attack' the pylon will have been sought beforehand. Using INAS or similar, the pilot will get himself to the start of his low-level run – the Initial Point. After this he goes visual, following the line on the map, and trying to keep to his planned timings. He flies below the hill summits where he can, using the terrain to mask his approach from the 'enemy radar defences'. In some modern aircraft types, moving maps and head-up displays make map-following easier.

Training sorties at low-level must be flown in good VMC and this restriction applies even to types such as the Tornado and F-111 which are equipped with TFR (Terrain Following Radar) capable of guiding the aircraft safely on a low-level run across any kind of terrain in darkness or bad weather.

Operational Roles

The principle of Flexible Response requires that the NATO air forces must be capable of a very varied repertoire of operational roles, and the large number of aircraft types still retained by the RAF and USAF reflect this need. At one end of the scale is the Hercules assault transport, hero of many an airlift and mercy mission, capable of going in low if necessary to avoid enemy radar and ground fire. At the other end of the scale are the Air Defence Squadrons – currently equipped with Phantoms (although some Lightnings still remain). Their specialised role is that of intercept-

045
2

XV

Figure 107 Control tower at RAF Wattisham. The observer, on the right, is a member of a Norwegian squadron on an exchange visit. *MOD*

Figure 108 Low-level flying is an important ingredient of RAF and USAF training. Here, a Buccaneer swoops over the Dortmund/Ems canal. *MOD*

Figure 109 Jaguars from five RAF squadrons based in Germany. *British Aerospace*

ing attackers, whether at high or at low level. Their specialised radar can look down from several thousand feet and pick out intruders hoping to sneak in low against the terrain. The Phantoms usually train in pairs, each practising interceptions on the other, high over the North Sea. They come under the direction of the Sector Operations Controllers who are responsible for the air battle from within their hardened bunkers near the air defence radar heads. Each Phantom has eight missiles, and as these are fired off one-by-one, hunter and quarry would approach close enough to each other to finish off the battle using guns.

Somewhere between the two extremes fall the other roles, from that of the mighty but mysterious Nimrod which patrols over the Atlantic hour after hour, and the ingenious Buccaneer, designed for the low-level anti-shipping strike role. For ground attack the RAF's Harrier, Jaguar and Tornado, and the USAF's F-111 can each boast some special characteristic. The Jaguar can use makeshift runways, whereas the VTOL Harrier needs no runway at all. Both are equipped with moving-map and head-up displays, and the Jaguars are very proud of their sophisticated navigation and weapon aiming system, plus a Laser-Marked Target Seeker. Both the Tornado and the F-111 have variable-geometry wings as well as advanced avionics. Both have Terrain Following Radar which can be used to fly low and fast in all weathers. The Flight Control System based on this radar automatically climbs the hills and descends into the valleys. In order to deliver an attack right on target at night or in cloud, the navigator uses his mapping radar and INAS in conjunction with one another. From a topographical map the navigator can predict how certain features along the run-in will show on radar, and can use these to update the INAS as he flies. The result is that the bombs can be delivered with pinpoint accuracy, darkness or heavy cloud notwithstanding.

It remains for all civilised people, general public and aviation enthusiasts alike, to hope and pray that these 'terrible swift swords' need never be used in anger.

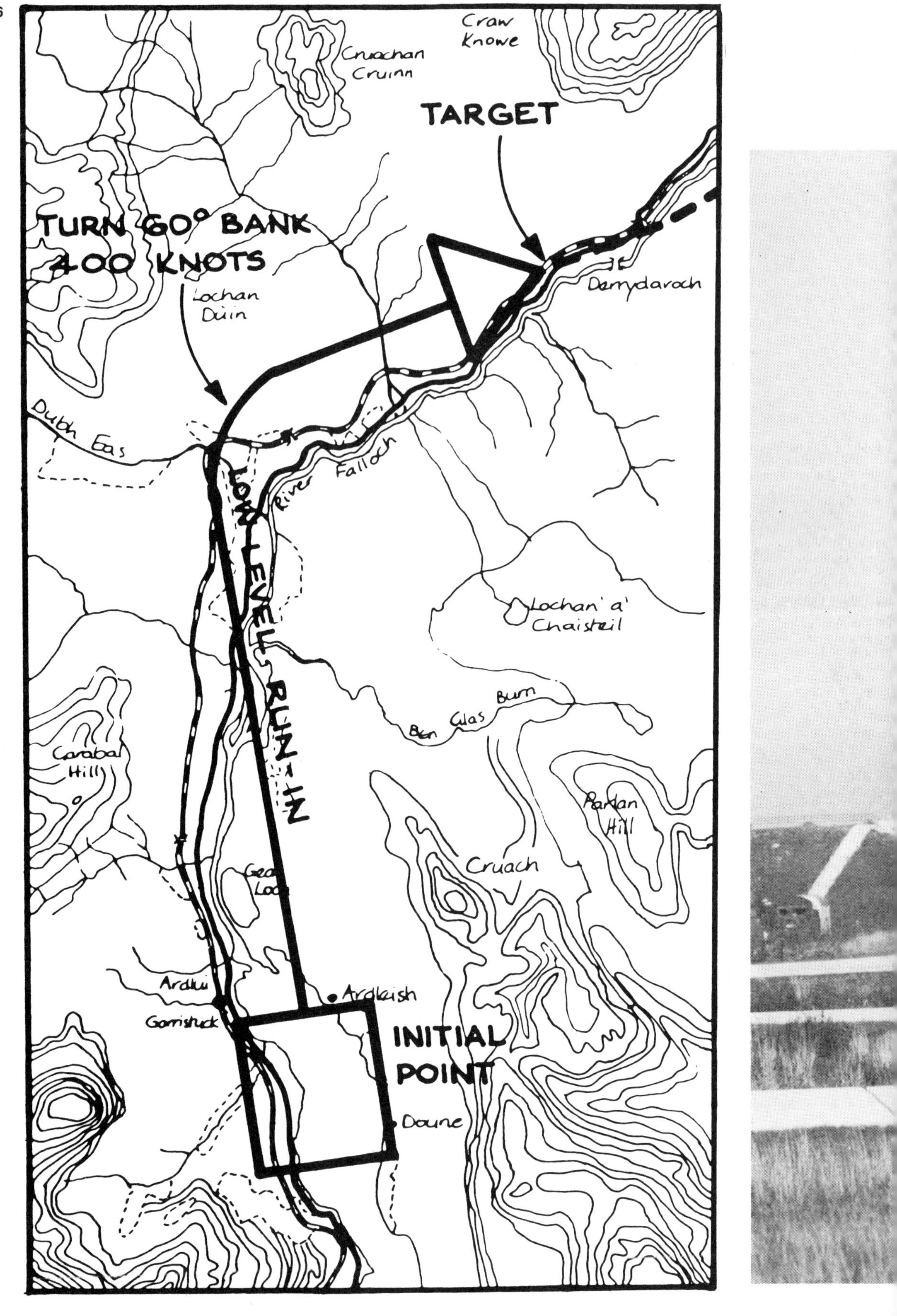
Craw Knowe
Cruachan Cruinn
TARGET
TURN 60° BANK
400 KNOTS
Lochan Dúin
Derrydaroch
Dubh Eas
River Falloch
LOW LEVEL RUN-IN
Lochan a' Chaisteil
Ben Glas Burn
Caraball Hill
Parlan Hill
Cruach
Ardlui
Ardleish
Garristuck
INITIAL POINT
Doune

Figure 110 Plan of a typical low-level training strike against a 'target' in the Scottish Highlands
Figure 111 A Jaguar operates from an unopened stretch of Autobahn near Bremen. *MOD*

Figure 112 Air defence. The Phantom carries underwing fuel tanks and four Sidewinder air-to-air missiles. On the ground are four Sparrow dogfight missiles and a centreline pod containing a 20 mm. rotary cannon. *MOD*

Figure 113 A Nimrod M.R. 1 maritime patrol aircraft. *MOD*

Figure 114 A number of squadrons of the American swing-wing F-111 all-weather bombers are now based in East Anglia. In this photo an F-111 formates with a KC-135 refuelling tanker. *USAF*

Figure 115 Boscombe Down based Harrier XV 281 launching from the 12° ski ramp at RAE Bedford. *MOD*

Index